RELATIVITY, GRAVITATION, COSMOLOGY

Contemporary Fundamental Physics
Valeri V. Dvoeglazov (Editor)

Relativity, Gravitation, Cosmology
Valeri V. Dvoeglazov and Augusto A. Espinoza Garrido
ISBN 1-59033-981-9

The Problem of Electron and Superluminal Signals
V.P. Olenik
ISBN 1-56072-938-4

The Photon
Trilochlan Pradhan
ISBN 1-56072-928-7

Symmetry and Electromagnetic Interactions of Fields with Multi-Spin
S. I. Kruglov
ISBN 1-56072-880-9

Photon and Old Problems in Light of New Ideas
Valeri V. Dvoeglazov
ISBN 1-56072-810-8

Photon and Poincaré Group
Valeri V. Dvoeglazov
ISBN 1-56072-718-7

Instantaneous Action at a Distance in Modern Physics: "Pro" and "Contra"
Andrew Chubykalo, Viv Pope and Roman Smirnov-Rueda
ISBN 1-56072-698-9

The Weyl-Dirac Theory and Our Universe
Mark Israelit
ISBN 1-56072-551-6

RELATIVITY, GRAVITATION, COSMOLOGY

VALERI V. DVOEGLAZOV
AND
AUGUSTO A. ESPINOZA GARRIDO
EDITORS

Nova Science Publishers, Inc.
New York

Production Coordinator: Susan Boriotti
Coordinating Editor/Graphics: Tatiana Shohov
Graphics: Levani Chlaidze
Senior Production Editors: Susan Boriotti and Donna Dennis
Production Editors: Marius Andronie and Rusudan Razmadze
Office Manager: Annette Hellinger
Editorial Production: Maya Columbus, Aloysha Klestov, Vladimir Klestov, Matthew Kozlowski and Lorna Loperfido
Circulation: Ave Maria Gonzalez, Vera Popovic, Luis Aviles, Jonathan Boriotti, Robert Brower, Alexandra Columbus, Raymond Davis, Cathy DeGregory, Melissa Diaz, Magdalena Nuñez, Marlene Nuñez, Jeannie Pappas and Frankie Punger
Communications and Acquisitions: Serge P. Shohov

Library of Congress Cataloging-in-Publication Data
Relativity, Gravitation, Cosmology / Valeri V. Dvoeglazov and Augusto A. Espinoza Garrido, editors.
p. cm. -- (Contemporary fundamental physics)
Includes index.
ISBN 1-59033-981-9 (hardcover)
1. Relativity (Physics) 2. Gravitation. 3. Cosmology. 1. Dvoeglazov, Valeri V. II Espinoza Garrido, Augusto A. III. Series.
QC173.58.R45 2004-06-09
530.11—dc22 2004005627

400 Oser Ave, Suite 1600
Hauppauge, New York 11788-3619
Tele. 631-231-7269 Fax 631-231-8175
e-mail: Novascience@earthlink.net
Web Site: http://www.novapublishers.com

Printed in the United States of America

Contents

EDITORIAL INTRODUCTION

In the end of 2002 we proposed to start a new Journal with the tentative title "Relativity, Gravitation, Cosmology" to be published by the Nova Science Publishers, NY, USA. The responses to this proposal were varied:

M. Carmeli: "Of course I will be interested in your program..."; Z. Oziewicz: "...Best wishes for your New and Important Journal "Relativity, Gravitation, Cosmology"; V. Simulik: "Very good project! Thank you for the information"; ***: "I am a co-editor of "Annalen der Physik" since 1 or 2 years. And this is certainly incompatible with being a co-editor of your journal. I am sure that you understand that"; R. Amoroso: "Congratulations on institution of new Gravity journal"; B. Sidharth: "Congratulations on launching this timely journal"; J. J. Slawianowski: "The idea you suggest looks very interesting"; L. E. Morales Guerrero: "Enhorabuena, Felicidades! Que gusto saber de algo asi"; J. M. Rivera Rebolledo: "Me parece interesante la propuesta"; ***: "I must inform you that I cannot support this project. I am strongly against new journals published by commercial publishers. With modern technology the science community is able to publish journals much less expensively and more effectively than can commercial publishers. I believe we should do everything we can to kill journals published commercially"; ***: "I suggest that rather than "launch" a new journal you simply go to lunch together and relax. There are already too many journals on the market. I am not interested in any news about this undertaking"; L. Horwitz: "I would be happy to serve on the Editorial Board if you wish. The topic has long been of interest to me".

We finally received 37 contributions for the first issue, among which we selected 14. So, in the first year we decided to publish an almanaque. We need, at least, 40 good papers per year, in order to justify marketing a new Journal. Meanwhile, we noted some attacks against us and at Physics by several persons in some Physical Forums and even in some Universities (2000-2004). In general, we doubt the claim that commercial Publishers cannot publish good things in Science, as well as that "Mexican Universities cannot be at the same scientific level as European or American Universities".

The volume consists of the 14 papers. We briefly describe their content:

- The R. Yamaleev paper presents a new extension of the relativistic mechanics based on the ideas, similar to the Kaluza and Rumer 5-dimensional theories.
- The theme is continued by Kociński and Wierzbicki. It is a nice review while not all the references to previous works have been given.
- Next we have another review, which proves the necessity of introduction of the Cartan torsion in the theory. R. Kühne argues that "the duality be-

tween curvature and torsion is analogous to the duality between electricity and magnetism".

- However, not all physicists agree with this viewpoint. While, one of the referees pointed out that he strongly disagrees with Garecki in his claims, but we decided to include the Garecki late contribution. Garecki is strongly against torsion in the gravitation theories.
- Tiwari returned to the old but beauty Weyl unified theory. What can we say? We can only ask: "Is he able to overcome objections to this idea which have been put forward by A. Einstein"?
- In order to characterize the Amoroso-Vigier paper we cite the referee expressions: "I still cannot understand the headline of this work and what new thing they actually do.". However, we still decided to publish this paper because we are convinced that we should present different viewpoints, even those which seem to us to not be in the mainstream.
- A very interesting paper by A. Camacho follows. Fluctuations of the space-time geometry due to quantum effects is a very exciting and timely topic.
- S. Ghosh advocates the approach in which the space-time non-commutativity should not be introduced ad-hoc. It should be derived from more underlying principles. We agree with him. Moreover, one of the editors (V. Dvoeglazov) published before some "speculative" contributions on this problem.
- L. Horwitz, our old friend, and his pupil O. Oron undertake a journey with the Brownian particles in the realtivistic world. Remember that it was A. Einstein who succeeded in the study of both topics.
- G.-J. Ni is brave as always to suggest revolutionary things. In the first paper he analyzed the "negative-mass paradox" in the gravitation theory, thus proposing a correction to the Newton law to include anti-gravity. In the second paper, he claimed to have explained "negative-mass paradox" in the neutrino physics, proposing a new scheme of minimal three-flavour neutrino oscillations. Some referee have been very exciting to read his papers. None found any errors.
- The Eganova paper had to be published in the volume "Instantaneous Action at a Distance in Modern Physics: Pro and Contra" of the series "Contemporary Fundamental Physics" of the Nova Science Publishers. However, somehow the paper was missed in that issue. In order to correct the error, V. Dvoeglazov (as an editor of the Series) decided to include the paper in this volume. This paper is based on the old and controversial Kozyrev's mechanics. Who knows, perhaps, scientists did not pay enough attention to it. Kozyrev himself claimed that his theory is proven by his own experiments. Some of them have been repeated; others have been rejected.

- The R. Kiehn paper return to the Cartan methods, now for the aims "to construct a cosmological model of the present universe". The responses of two referees were rather positive. We leave to decide to readers about the merits of this paper.
- The R. Cahill paper concludes the volume. It is interesting by its historical discussions, particularly, of experiments designed to detect absolute motion. Moreover, the paper presents a new theory of gravitation.

We sincerely hope that these papers will be useful to our readers. We wish good luck to our Publisher, our authors and our critics. We wait good papers. We are going to continue the publication of books/almanaques/journals on the announced topics of Relativity, Gravitation and Cosmology in the Nova Science Publishers. We have full support of the Publisher. So, the announcement for the next volume is the same as before:

Announcement

The Nova Science Publishers (NY, USA) decided to launch a new Project of the Journal "Relativity, Gravitation, Cosmology". We want to start this Journal with special issues dedicated to the following themes:

- Dilaton gravity
- Quantum Mechanical Phases, Neutrino and Gravity. Photon and Gravity
- Spin connection and 4-potential. Axion, Torsion and Notoph
- Curvature as a Scalar Field over the Minkowski Space
- Multidimensional Gravity. De Sitter Gravity. Weyl Approach
- Relativistic Quantum Mechanics Approach to Gravity (a la S. Weinberg). Parity Violation
- Non-commutative Space-time.

The papers should be submitted to Prof. V. Dvoeglazov (valeri@ahobon.reduaz.mx). We hope that the Gravity community will approve the new Journal and will submit papers on the permanent basis. It is assumed that 4 issues per year will be published. The acceptable topics of papers are indicated in the title of the Journal and are not restricted by the themes of the first special issue.

We are ready to consider other candidates for the Editorial Board of the Journal.

Inquiries concerning submissions of papers and subscriptions to the Journal should be sent to donnanova@earthlink.net or valeri@ahobon.reduaz.mx and novascience@earthlink.net.

Valeri V. Dvoeglazov
Augusto A. Espinoza Garrido

Universidad de Zacatecas
Apartado Postal 636, Suc. UAZ
Zacatecas 98062 Zac.
México

In: Relativity, Gravitation, Cosmology
Editor: V. Dvoeglazov and A. Espinoza Garrido, pp. 1-17
ISBN 1-59033-981-9

Chapter 1

Extended Relativistic Mechanics of Charged Particle

Robert M. Yamaleev
FES-C UNAM, Fisica, Cuautitlán Izcalli, C.P.54740, México
141980 JINR, LIT, Russia
e-mail:iamaleev@servidor.unam.mx

Abstract

We build extension of the relativistic mechanics. The extension is carried out by introducing a new component of the momentum and a new evolution parameter in such a way that the energy conservation law is transformed into one of the dynamic equations of the extended relativistic mechanics. The extended Lorentz-force equations imply two constants of motion which are proportional to the mass and the charge of a particle, correspondingly. The celebrated Einstein's formula $E = mc^2$ is generalized. The corresponding Hamilton and Hamilton-Jacobi equations are derived. Modified Schrödinger and Dirac equations are formulated.

1.1 Introduction

In order to unify the gravitation with the electromagnetism T.Kaluza [1] has introduced a fifth component of the momentum related with the charge of the particle. Later this idea has been developed by many authors (see, [2] and references therein). One of the fundamental developments of the Kaluza's idea has been done by Yu.Rumer [3]. In these works, influences of an electric charge on a space-time are treated as sources of curvature in a similar way as mass does in

Einstein's theory of the general relativity. Thus, the main subject of investigation has been an extension of Einstein's equations which must contain the original Einstein's equations and Maxwell's equations. The corresponding geodesic equations in five dimensional space-time must contain the original geodesic equations and the Lorentz-force equations. As it has been noted by Rumer, these equations are written with respect to the new parameter of evolution, not the proper-time. Rumer interpreted this parameter as a function of action.

In the present paper we use the Kaluza's idea to extend the classical relativistic dynamics. Our approach distinguishes from the approaches of the followers of the Kaluza's idea by the following: we don't touch the relativistic quadratic form connecting four-momentum and mass; we add a new quadratic relationship connecting the new component of the momentum with the charge. We start from the observation that the relativistic extension of the Lorentz-force equation is fulfilled in such a way that the energy conservation law is transformed into one of the dynamic equations. This procedure of extension we formulate as an algorithm and use together with the Kaluza's idea in order to extend the relativistic mechanics. The extended Lorenz-force equations imply two constants of motion, the first one is identified with the inertial mass, and the second, with the charge of a particle. This extension needs a new scale parameter in the units of the ratio of mass by charge.

The paper is organized as follows:

In Sec. 2, we briefly review some features of the relativistic dynamics of the charged particle moving inside external electromagnetic fields. Equations of motion are given by Lorentz-force equations. The pathway from three-vector representation into four-vector representation is formulated as an algorithm. It is shown, on the direction of the momentum the Lorentz-force equations are reduced into equations for hyperbolic cosine-sine functions. A new method of transition from Lorentz-force equations into Hamilton-Jacobi equation is presented.

In Sec. 3, by using the algorithm elaborated in Sec. 2 we extend Lorentz-force equations. A new scale factor in the units of the ratio of mass by charge is introduced. We define also a dimensionless constant $1 \geq k \geq 0$. In the limit $k = 0$, the extended equations are reduced to the relativistic Lorentz-force equations. On the direction of the momentum the extended Lorentz-force equations are given by equations for Jacobi elliptic functions. In this context, the presented extension can be understood as a deformation hyperbolic cosine-sine functions into Jacobi elliptic functions, where k plays the role of the parameter of deformation. The equations possess three contractions: a) $k = 0$, b) $k = 1$, c) $k = -1$.

In Sec. 4, a motion of the charged particle under a static external field is investigated.

In Sec. 5, the Hamilton and Hamilton-Jacobi equations for the extended Lorentz-force equations are derived.

In Sec. 6, the modified Schrödinger, Klein-Gordon and Dirac equations are formulated.

The paper closes with the Summary.

1.2 The Lorentz-force equations

Due to the concept of the special relativity a greatly advance has been accomplished in such fields of the physics as gravitation, quantum mechanics and quantum electrodynamics. Sensible development underwent the classical mechanics, specially, the dynamics of the charged particles. The flow of charged particles serves as a source of the electromagnetic fields $\vec{E}$, $\vec{B}$ and, on the other hand, the electromagnetic fields affect on a motion of the charged particles with the force defined by H.A.Lorentz [4]

$$\frac{d\vec{p}}{dt} = Ze(\vec{E} + [\vec{v} \times \vec{B}]), \tag{2.1a}$$

where $\vec{E}$ is the electric field strength and $\vec{B}$ is the magnetic flux density. These fields are related with the potentials $(\Phi,\ \vec{A})$ by

$$\vec{E} = -\vec{\nabla}\Phi - \frac{\partial \vec{A}}{\partial t}, \quad \vec{B} = [\vec{\nabla} \times \vec{A}]. \tag{2.2}$$

Eqs.(2.1a) must be complemented with a connection between the momentum and the velocity. The relativistic correction concerns this connection. Newtonian mechanics implies the relation

$$\vec{p} = m\vec{v}. \tag{2.1b}$$

In the relativistic mechanics this connection is given by more complicate function

$$\vec{p} = \frac{m\vec{v}}{\sqrt{1 - \frac{v^2}{c^2}}}. \tag{2.1c}$$

In the sequel, founders of the relativistic mechanics H.Poincaré [5], A.Einstein [6], M.Planck [7] and Minkowski [8] showed that the Lorentz-covariance of the Lorentz-force equation is provided by introducing the fourth component of the momentum. These equations had been formulated with respect to the invariant evolution parameter, proper time. The textbook presentations usually demonstrate a pass from four-dimensional formulation into three-dimensional one. For our purpose it is useful to pass this pathway in the opposite direction.

The three-dimensional relativistic Lorentz-force equations (2.1a), (2.1c) can be cast into four-dimensional formulation by using the following
Algorithm.
From Eq.(2.1a) we obtain an evolution equation for the kinetic energy

$$\frac{d}{dt}\frac{p^2}{2m} = Ze(\vec{E} \cdot \vec{p})\frac{1}{m}. \tag{2.3}$$

Define a kinetic-like energy which differs of the kinetic energy by a constant of motion:

$$\frac{p_0^2}{2m} = \frac{p^2}{2m} + C. \tag{2.4}$$

The new kinetic energy, obviously, obeys the same evolution equation

$$\frac{d}{dt}\frac{p_0^2}{2m} = Ze(\vec{E}\cdot\vec{p})\frac{1}{m}. \tag{2.5}$$

Define a new evolution parameter τ by

$$\frac{p_0}{mc} = \frac{dt}{d\tau}. \tag{2.6}$$

In the sequel, τ has been interpreted as the proper-time of the particle. With respect to the proper-time equations (2.1a), (2.1c) and (2.5) are written as

$$\frac{d\vec{p}}{d\tau} = Ze(\vec{E}\,\frac{p_0}{mc} + [\vec{u}\times\vec{B}]), \tag{2.7a}$$

$$\vec{u} = \frac{d\vec{r}}{d\tau} = \frac{\vec{p}}{m}, \tag{2.7b}$$

$$\frac{dp_0}{d\tau} = Ze(\vec{E}\cdot\vec{p})\frac{1}{mc}. \tag{2.7c}$$

From these equations we find the first integral

$$p_0^2 - p^2 = M^2c^2. \tag{2.8}$$

The constant of motion M is equal m for a *massive* particle and is equal to zero for a *massless* particle. Notice, that for the massless particle the equality $M = 0$ does not necessitate the condition $m = 0$. Indeed, an interpretation of M as a mass of the particle comes from the principle of correspondence with the Newtonian mechanics. When one takes $M = 0$ this correspondence breaks down. The matter is that for the massless particle (with $M = 0$) the parameter m as well as the proper time τ are undefined. However, if one defines m then τ can be defined, too. And vice versa, a definition of τ means that m can be also defined.

In the case of stationary potential field $\Phi(r)$ another integral of motion is found

$$cp_0 + Ze\Phi(r) = \mathcal{E}_0. \tag{2.9}$$

In the direction of the vector of momentum the Lorentz-force equations are represented by the pair of equations [9]:

$$\frac{dp}{d\phi} = p_0, \quad \frac{dp_0}{d\phi} = p, \tag{2.11}$$

with the solutions

$$p_0 = mc\,cosh(\phi), \quad p = mc\,sinh(\phi), \tag{2.12}$$

where ϕ obeys the Newtonian equation

$$\frac{d\phi}{d\tau} = \frac{Ze}{mc}(\vec{E}\cdot\vec{n}), \quad \vec{n} = \frac{\vec{p}}{p}. \tag{2.13}$$

This form of Lorentz-force equations we call *polar representation.*

Hamilton-Jacobi equation

The Hamilton-Jacobi equation is an equation for the Lorentz-invariant function $S(r,t)$ called as the *function of action.* Let us outline a new method of construction of Hamilton-Jacobi equation from the Lorentz-force equations. In the sequel, we shall use this method to obtain Hamilton-Jacobi equation from the Extended Lorentz-force equations.

The following theorem holds true:

Theorem

Suppose that the components of the momentum $\{p, p_0\}$ are functions of the coordinate time t, as well of the spatial coordinates $\vec{r}$ and obey the relationship

$$p_0 = \sqrt{p^2 + m^2c^2}. \tag{2.14}$$

Then within the definitions (2.2) and

$$cp_0 = -\frac{\partial S}{\partial t} - Ze\Phi, \quad \vec{p} = \vec{\nabla}S - Ze\vec{A}, \tag{2.15}$$

the Lorentz-force equations (1.1a,c) are transformed into an identity.

Proof

From formulae (2.15) one finds

$$\Phi = -\frac{1}{Ze}(cp_0 + \frac{\partial S}{\partial t}), \quad \vec{A} = -\frac{1}{Ze}(\vec{p} - \vec{\nabla}S). \tag{2.16}$$

Then by substituting these potentials into (2.2) one gets

$$\vec{E} = \frac{1}{Ze}(\frac{\partial \vec{p}}{\partial t} + \vec{\nabla}cp_0), \quad \vec{B} = -\frac{1}{Ze}[\vec{\nabla}\times\vec{p}]. \tag{2.17}$$

Taking into account (2.14), the expression for $\vec{E}$ is written as

$$\frac{\partial \vec{p}}{\partial t} + \vec{\nabla}c\sqrt{p^2 + m^2c^2} = Ze\vec{E}. \tag{2.18}$$

Now consider the term $[\vec{v}\times\vec{B}]$. By using the formula

$$\vec{v} = \frac{c}{\sqrt{p^2 + m^2c^2}}\vec{p} \tag{2.19}$$

and the formula for $\vec{B}$, one obtains

$$Ze[\vec{v}\times\vec{B}] = -\frac{c}{\sqrt{p^2 + m^2c^2}}[\vec{p}\times[\vec{\nabla}\times\vec{p}]]. \tag{2.20}$$

On the other hand, the full derivative of the momentum with respect to the coordinate time is

$$\frac{d}{dt}\vec{p} = \frac{\partial}{\partial t}\vec{p} + (\vec{v}\cdot\vec{\nabla})\vec{p} \tag{2.21}$$

Substitute here $\vec{v}$ from (2.19) and use the formula

$$(\vec{p}\cdot\vec{\nabla})\vec{p} = \frac{1}{2}\vec{\nabla}(p^2) - [\vec{p}\times[\vec{\nabla}\times\vec{p}]]. \tag{2.22}$$

This gives

$$\frac{d\vec{p}}{dt} = \frac{\partial\vec{p}}{\partial t} + \vec{\nabla}c\sqrt{p^2+m^2c^2} - \frac{c}{\sqrt{p^2+m^2c^2}}[\vec{p}\times[\vec{\nabla}\times\vec{p}]],$$

which, according to formulae (2.18) and (2.20), is the sum of the terms $Ze\vec{E}$ and $Ze[\vec{v}\times\vec{B}]$.

End of proof

By introducing the definitions (2.15) into relationship (2.14) we obtain the relativistic Hamilton-Jacobi equation

$$-\frac{\partial S}{\partial t} - Ze\Phi = \sqrt{(\vec{\nabla}S - Ze\vec{A})^2 + m^2c^2}. \tag{2.23}$$

Notice that the non-relativistic connection between the velocity and the momentum implies the following definition for p_0:

$$p_0 = \frac{p^2}{2mc}.$$

1.3 Extended Relativistic Mechanics of Charged Particle

In Refs.[10], [11] a new way of generalization of the relativistic mechanics has been done. In polar representation the extended equations of motion were represented by Weierstrass equations for elliptic functions. Now we present the other approach which in the polar representation leads to Jacobi equations for elliptic functions.

Let us start from the relativistic Lorentz-force equation (2.7a)

$$\frac{d\vec{p}}{d\tau} = Ze(\vec{E}\,\frac{p_0}{mc} + [\vec{u}\times\vec{B}]), \tag{3.1}$$

where τ in the relativistic kinematics is interpreted as a proper-time of the particle. From these equations the following evolution equation for the kinetic energy is derived

$$\frac{d}{d\tau}\frac{p^2}{2m} = \frac{Ze}{m}(\vec{E}\cdot\vec{p})\,\frac{p_0}{mc}. \tag{3.2}$$

Thus, the Lorentz-force equations one may extend in a similar manner what it has been done in Sections 2. Firstly, let us define a *new* kinetic energy which differs from the original one by a constant of motion:

$$\frac{p_1^2}{2m_1} = \frac{p^2}{2m} + C, \tag{3.3}$$

where dim(p_1)=dim(p), dim(m_1)=dim(m).
The *new* kinetic energy satisfies the same equation as the *old* one. Therefore

$$\frac{d}{d\tau}\frac{p_1^2}{2m_1} = \frac{Ze}{m}(\vec{E}\cdot\vec{p})\,\frac{p_0}{mc}. \tag{3.4}$$

Secondly, define a new evolution parameter s by

$$\frac{d\tau}{ds} = \frac{p_1}{m_1 c}, \tag{3.5}$$

and re-write Eqs.(2.7a,b,c) and (3.4) with respect to the new parameter of evolution. We obtain the following system of equations

$$\frac{d\vec{p}}{ds} = Ze(\vec{E}\,\frac{p_0}{mc}\,\frac{p_1}{m_1 c} + [\vec{w}\times\vec{B}]), \tag{3.6a}$$

$$\frac{dp_0}{ds} = \frac{Ze}{mc}(\vec{E}\cdot\vec{p})\,\frac{p_1}{m_1 c}, \quad \frac{dp_1}{ds} = \frac{Ze}{mc}(\vec{E}\cdot\vec{p})\,\frac{p_0}{mc}. \tag{3.6b}$$

Here the velocity $\vec{w}$ is defined with respect to s:

$$\vec{w} := \frac{d\vec{r}}{ds}, \text{ and } \vec{p} = m\vec{w}. \tag{3.7}$$

The obtained system of equations we shall denominate by "*Extended Relativistic Lorentz-force Equations*". These equations imply two constants of motion

$$M^2c^2 = p_0^2 - p^2, \quad M_1^2c^2 = p_1^2 - \frac{m_1}{m}p^2. \tag{3.8}$$

The first one is nothing else than the relativistic mass-shell equation. Hence, the constant of motion M in the first quadratic form is equal to the inertial mass of the particle m. Notice, from this relationship it follows that at the rest the fourth component of the momentum p_0 is equal to mc.

Now, let us formulate *polar representation* for the Extended Lorentz-force equations. For that purpose consider the projection of Eqs.(3.6a,b) on the direction of the vector of momentum. Then Eqs.(3.6a,b) can be written as follows

$$\frac{d}{d\phi}p = p_0\,\frac{p_1}{m_1 c}, \quad \frac{d}{d\phi}p_0 = p\,\frac{p_1}{m_1 c}, \quad \frac{d}{d\phi}p_1 = p\,\frac{p_0}{mc}, \tag{3.10a}$$

where the angular parameter ϕ obeys the equation

$$\frac{d\phi}{ds} = \frac{Ze}{mc}\,(\vec{E}\cdot\vec{n}), \quad \vec{n} = \frac{\vec{p}}{p}. \tag{3.10b}$$

The solutions of Eqs.(3.10a) are given by Jacobi elliptic functions of imaginary argument (see, Eq.(A8) in Sec. 8 “Appendix”):

$$p = mc\ sc(\phi, k), \quad p_0 = mc\ nc(\phi, k), \quad p_1 = m_1 c\ dc(\phi, k) \text{ with } k = \frac{m}{m_1}. \quad (3.11)$$

At the rest we have $p_1 = m_1 c$. Hence, the constant of motion in the second quadratic form of (3.8) is

$$M_1 = m_1.$$

Interpretation

The following interpretation of the second mass parameter m_1 can be done. The charged particle has two physical characteristics: the mass and the charge. The mass we have used for interpretation the first constant of motion. Hence, the charge we may use in order to interpret the second constant of motion. Suppose that m_1 is proportional to the charge of the particle:

$$m_1 = Ze\ K,$$

where the coefficient K has the unit of a ratio of mass by charge. This ratio one may form by using two fundamental constants of the electromagnetism and the gravitation, namely, the *permittivity*, or the dielectric constant of evacuated (free) space ϵ_0 and the gravitational constant γ:

$$K := \frac{1}{\sqrt{\gamma\epsilon_0}}. \quad (3.12)$$

This is only one of the possibilities. Experimental measurements can give the other value for K. As an example let us mention the situation with the Planck's constant h. On making use constants of the electromagnetism, namely, the electric permittivity ϵ_0, the magnetic permeability μ_0 and the charge e one may form a constant possessing the same unit as the Planck's constant h. This constant is given by

$$h_e = e^2 \sqrt{\frac{\mu_0}{\epsilon_0}}.$$

Compare the value of this constant $h_e = 0.9669\ 10^{-35}[\frac{kg\ m^2}{sec}]$ with the value of the Planck's constant

$$h = 6.6256\ 10^{-34}[\frac{kg\ m^2}{sec}].$$

One half of the ratio of these constants is the Sommerfeld's *fine structure* constant α:

$$\alpha = \frac{1}{2}\frac{h_e}{h} = \frac{1}{137.043048}.$$

Contractions

We have seen that the components of the momentum are represented by Jacobi elliptic functions. These functions are double periodic and have three

contractions. The first one is obtained by equating the parameter k to 0. In that case the functions $sc(\phi,k),\ nc(\phi,k),\ dc(\phi,k)$ are reduced to hyperbolic cosine-sine functions:
$sc(\phi,0)=sinh(\phi),\ nc(\phi,0)=cosh(\phi),\ dc(\phi,0)=1.$
In this limit Eqs.(3.6a,b) are reduced to the relativistic Lorentz-force equations. The second limit is obtained by taking $k=1$ which corresponds to the case $m=m_1$. In this case the functions $sc(\phi,k),\ nc(\phi,k),\ dc(\phi,k)$ are transformed into the periodic trigonometric functions:

$$sc(\phi,1)=tan(\phi),\quad nc(\phi,1)=dc(\phi,1)=\frac{1}{cos(\phi)}.$$

The case $k=-1$ is of a special interest for further investigations. In this case the Jacobi elliptic functions are represented by *lemniscate function* [12].

The coordinate time as an evolution parameter

The connection between invariant time s and the coordinate time t is defined from (2.6) and (3.5):

$$\frac{p_0p_1}{mm_1c^2}=\frac{dt}{ds} \tag{3.13}$$

So, in order to rewrite Eqs.(3.6a,b) with respect to the coordinate time t we should use the following relationship between the derivatives

$$\frac{d}{ds}=\frac{p_0p_1}{mm_1c^2}\frac{d}{dt}. \tag{3.14}$$

Then the Extended Lorentz-force equations take the following form

$$\frac{d\vec{p}}{dt}=Ze(\vec{E}+[\vec{v}\times\vec{B}]), \tag{3.15a}$$

$$p_0\frac{dp_0}{dt}=Ze(\vec{E}\cdot\vec{p}),\quad p_1\frac{dp_1}{dt}=Ze\,\frac{m_1}{m}(\vec{E}\cdot\vec{p}), \tag{3.15b}$$

where the connection between the momentum and the velocity is given by

$$\vec{v}:=\frac{dr}{dt}=\vec{p}\frac{\sqrt{mm_1}c^2}{\sqrt{p^2+m^2c^2}\sqrt{p^2+mm_1c^2}}. \tag{3.16}$$

From this formula one may derive the function $p=p(v)$. For that purpose one has to solve the quadratic equation

$$X^2+X\ mc^2(m+m_1(1-\frac{c^2}{v^2}))+m^3m_1c^4=0,\ \text{with}\ X=p^2. \tag{3.17}$$

This equation has two solutions with different behaviours at the initial point $v=0$:
$x_1(v=0)=0,\ x_2(v=0)=\infty.$
Among two solutions we have to choose one what corresponds to the normal

behaviour at the rest, that is $p^2 = x_1$.
This gives the following dependence p^2 of the velocity v:

$$p^2 = -\frac{mc^2}{2}(m + m_1(1 - \frac{c^2}{v^2}) - \sqrt{(m + m_1(1 - \frac{c^2}{v^2}))^2 - 4mm_1} \tag{3.18}$$

In the limit $m_1 \to \infty$ this formula tends to the relativistic relationship between squares of the momentum and the velocity

$$p^2 = \frac{m^2 v^2}{1 - \frac{v^2}{c^2}}. \tag{3.19}$$

1.4 Motion of charged particle under a static electric field

Consider one-dimensional motion under the static electric field $\Phi(r)$. This motion is described by the equations

$$\frac{dp}{ds} = -Ze\frac{d\Phi}{dr}\frac{p_0}{mc}\frac{p_1}{m_1 c}, \quad m\frac{dr}{ds} = p, \tag{4.1a}$$

$$\frac{dp_0}{ds} = -Ze\frac{d\Phi}{dr}\frac{p}{mc}\frac{p_1}{m_1 c}, \quad \frac{dp_1}{ds} = -Ze\frac{d\Phi}{dr}\frac{p}{mc}\frac{p_0}{mc}. \tag{4.1b}$$

From these equations we find

$$p_0 = c\sqrt{mm_1 - m^2}\,\sinh(R), \quad p_1 = c\sqrt{m_1^2 - m_1 m}\,\cosh(R), \tag{4.2}$$

where

$$R := \sqrt{\frac{m}{m_1}}\frac{E_0 - Ze\Phi}{mc^2}.$$

The momenta p_0, p, p_1 obey the following relationships:

$$p_0 = \sqrt{p^2 + m^2c^2}, \ p_1 = \sqrt{\frac{m_1}{m}p^2 + m_1^2c^2}. \tag{4.3}$$

From (4.2) and (4.3) it follows

$$\tanh\left(\sqrt{\frac{m}{m_1}}\frac{E_0 - Ze\Phi}{mc^2}\right) = \sqrt{\frac{p^2 + m^2c^2}{p^2 + mm_1c^2}}.$$

By inverting the left-hand side we obtain the following expression for the energy

$$E_0 = \sqrt{mm_1}c^2 \tanh^{-1}\left(\sqrt{\frac{p^2 + m^2c^2}{p^2 + mm_1c^2}}\right) + Ze\Phi. \tag{4.4a}$$

This formula can be written also in the next form

$$E_0 = \frac{1}{2}\sqrt{mm_1}c^2\cosh^{-1}(1+2\frac{p^2+m^2c^2}{m(m_1-m)c^2}) + Ze\Phi. \qquad (4.4b)$$

The rest energy formula

The expression for the rest energy one obtains from formulae (4.4a,b) by taking $p=0$ and $\Phi=0$. One gets

$$E_0 = \frac{1}{2}\sqrt{mm_1}c^2\cosh^{-1}(\frac{m_1+m}{m_1-m}). \qquad (4.5a)$$

This formula can be written in the other forms, too. For instance,

$$E_0 = \sqrt{mm_1}c^2\tanh^{-1}(\sqrt{\frac{m}{m_1}}), \text{ or } E_0 = \frac{\sqrt{mm_1}c^2}{2}\log\frac{1+\frac{m}{m_1}}{1-\frac{m}{m_1}}. \qquad (4.5b)$$

These formulas tend to celebrated Einstein's formula $E_0 = mc^2$ when $k = \frac{m}{m_1} \to 0$.

1.5 Hamilton and Hamilton-Jacobi equations

Hamiltonian $H(r,p)$ for the Extended Relativistic Lorentz equations (3.15a,b) is defined by

$$H = \frac{1}{2}\sqrt{mm_1}c^2\cosh^{-1}(1+2\frac{p^2+m^2c^2}{m(m_1-m)c^2}) + Ze\Phi(r). \qquad (5.1)$$

By inserting this function into the Hamilton equations

$$\frac{d}{dt}\vec{p} = -\frac{\partial H}{\partial \vec{r}}, \quad \frac{d}{dt}\vec{r} = \frac{\partial H}{\partial \vec{p}} \qquad (5.2)$$

we obtain Eqs.(3.15a,b).

The Hamilton-Jacobi equation corresponding to Eqs.(3.15a,b) is constructed by applying the method given in Section 2. In the case of Extended Lorentz-force equations the connection between the momentum and the velocity is defined by (3.16). In that case we have to use to the following formula

$$\vec{\nabla}p_0 = \frac{1}{2}\sqrt{mm_1}c^2\frac{1}{\sqrt{(p^2+m^2c^2)(p^2+mm_1c^2)}}\vec{\nabla}p^2.$$

Correspondingly, the relationship between p_0 and p has the form

$$p_0 = \frac{1}{2}\sqrt{mm_1}c\ cosh^{-1}(1+2\frac{p^2+m^2c^2}{m(m_1-m)c^2}).$$

By using this formula instead of (2.14) we arrive to the following Hamilton-Jacobi equation

$$\frac{m(m_1-m)c^2}{2}\ cosh(2R) = p^2 + \frac{m(m+m_1)c^2}{2}, \qquad (5.3)$$

where

$$R = \frac{1}{\sqrt{mm_1}c^2}(-\frac{\partial S}{\partial t} - Ze\Phi(r)), \quad \text{and } \vec{p} = \vec{\nabla}S - Ze\vec{A}.$$

Formula (5.3) can be written also as follows

$$m(m_1 - m)c^2 \sinh^2(R) - p^2 = m^2c^2. \tag{5.4}$$

It is easily seen that the Hamilton-Jacobi equations (5.3)-(5.4) tend to the relativistic one when $m_1 \to \infty$, or $k = \frac{m}{m_1} \to 0$.

1.6 Modified Schrodinger and Dirac equations

Within the framework of Schrodinger scheme of quantization the derivatives of the action are replaced by the corresponding differential operators according to the rule:

$$\frac{\partial S}{\partial t} \to -i\hbar\frac{\partial}{\partial t}, \quad \vec{\nabla}S \to -i\hbar\vec{\nabla}. \tag{6.1}$$

By replacing derivatives of the action in the Hamilton-Jacobi equation (5.3) by the corresponding differential operators, we get the Schrodinger-like operator acting on the wave function Ψ:

$$\frac{m(m_1 - m)c^2}{2}\cosh(2\Xi)\ \Psi = (p^2 + \frac{m(m_1 + m)c^2}{2})\ \Psi, \tag{6.2}$$

where

$$\vec{p} := -i\hbar\frac{\partial}{\partial \vec{r}} - Ze\vec{A}, \text{ and } \Xi := \frac{1}{\sqrt{mm_1}c^2}(i\hbar\frac{\partial}{\partial t} - Ze\Phi)). \tag{6.3}$$

The operator of the boost along t acts according to Euler's formula:

$$\exp(i\frac{\hbar}{\sqrt{mm_1}c^2}\frac{\partial}{\partial t})\Psi(t, r) = \Psi(t + i\frac{\hbar}{\sqrt{mm_1}c^2},\ r). \tag{6.4}$$

Hence,

$$e^{i\frac{\hbar}{\sqrt{mm_1}c^2}\frac{\partial}{\partial t}}e^{-i\frac{Et}{\hbar}} = e^{-i\frac{Et}{\hbar}}e^{\frac{2E}{\sqrt{mm_1}c^2}}. \tag{6.5}$$

Thus, with respect to the coordinate time we have difference-differential equation. In this context let us mention the *relativistic quasipotential equation* of Kadyshevsky-Mir-Kasimov-Skachkov [15] , which contains difference-differential operator acting on the coordinate space.

In order to obtain modified Klein-Gordon equation we should start from (5.4). Exchanging the derivatives of the action in (5.4) by the corresponding differential operators according to the rule (6.1), we get modified Klein-Gordon equation

$$(c\sqrt{m(m_1 - m)})^2\ sinh^2(\Xi)\ \Psi = (p^2 + m^2c^2)\ \Psi. \tag{6.4}$$

Linearize this equation in the basis of the Clifford algebra. The algebraic procedure of linearization is carried out by ordinary way. This gives modified Dirac equation:

$$(\, c\sqrt{m_1 m - m^2}\; sinh(\Xi)\; \gamma_0 + p^k \gamma_k \;)\mathcal{X} = mc\; \mathcal{X}, \tag{6.5}$$

where $\mathcal{X}$ is Dirac's bispinor, γ_μ, $\mu = 0, 1, 2, 3$ are Dirac's *gamma*-matrices.

1.7 Summary

We proposed an extension of the relativistic dynamics of the charged particle. The algorithm of the extension consists of two stages:
1) definition a new component of the momentum;
2) definition a new evolution parameter.
In this way the equation for the kinetic energy is involved into the system of dynamic equations. The Lorentz-force equation and it's extension permit, so called, polar representation where the components of the momentum are expressed via the angular parameter obeying the Newtonian-like equation. In the polar representation the momenta of the relativistic Lorentz-force equations are given by the set of hyperbolic cosine-sine functions. For the extended Lorentz-force equations the components of the momentum are given by the set of Jacobi elliptic functions of imaginary argument. Further extensions, obviously, lead to the set of hyperelliptic functions. In this hierarchy we have restricted ourselves only with the Jacobi elliptic functions. For these functions still addition theorem is valid. This theorem is important to construct corresponding generalization of the Lorentz kinematics.

The relativistic Lorentz-force equation in four-vector formulation implies the first integral of motion given by the quadratic relationship between the components of the four-vector of momentum. This relationship admits interpretation of the fourth component via the inertial mass of the particle. The extended Lorentz-force equations imply an additional integral of motion which restricts the fifth component of the momentum. The unique alternate candidate to interpret this new variable is the charge of the particle. The theory contains a universal constant of dimensionality of the ratio of mass by charge.

By using a new method of deriving of Hamilton-Jacobi equations we came to the modified Hamilton-Jacobi equation and its quantum analogues — the modified Schrodinger and Klein-Gordon equations. From the last by using ordinary procedure of algebraic linearization we came to the modified Dirac equation.

1.8 Appendix. Jacobi elliptic function

The elliptic function are a natural extension of the circular functions [13]. The most important property of the elliptic functions is double periodicity, in contrast to single periodicity of the circular functions.

Jacobi elliptic functions
$sn(u,k),\ cn(u,k),\ dn(u,k)$ obey the following differential equations

$$\frac{d}{du}sn(u,k) = cn(u,k)dn(u,k),$$

$$\frac{d}{du}cn(u,k) = -sn(u,k)dn(u,k),$$

$$\frac{d}{du}dn(u,k) = -k\ cn(u,k)sn(u,k), \tag{A.1}$$

with the initial conditions

$$sn(0,k) = 0,\ cn(0,k) = 1,\ dn(0,k) = 1. \tag{A.2}$$

From (A.1), (A.2) the following quadratic relations are deduced

$$cn^2(u,k) + sn^2(u,k) = 1, \quad dn^2(u,k) + k\ sn^2(u,k) = 1. \tag{A.3}$$

When $k \to 0$ the Jacobi elliptic functions give arise sine and cosine functions:

$$sn(u,0) = \sin(u),\ cn(u,0) = \cos(u),\ dn(u,0) = 1.$$

When $k = 1$, $cn(u,1) = dn(u,1)$. In this limit the following correspondence holds:

$$sn(u,1) = tanh(u),\ cn(u,1) = \frac{1}{\cosh(u)},\ dn(u,1) = \frac{1}{\cosh(u)}.$$

Jacobi's notation for the functions $sn(u),\ cn(u),\ dn(u)$ was $sinam(u)$, $cosam(u)$, $\Delta am(u)$, the abbreviations now in use being due to Gudermann [14], who also wrote $tn(u)$, as an abbreviation for $tanam(u)$, in place of what is now written $sc(u)$.

Glaisher's notation for quotients
A short and convenient notation has been invented by Glaisher to express reciprocals and quotients of the Jacobian elliptic functions; the reciprocals are denoted by reversing the order of the letters which express the function, thus

$$ns(u) = \frac{1}{sn(u)},\quad nc(u) = \frac{1}{cn(u)},\quad nd(u) = \frac{1}{dn(u)},$$

$$sc(u) = \frac{sn(u)}{cn(u)},\quad sd(u) = \frac{sn(u)}{dn(u)},\quad cd(u) = \frac{cn(u)}{dn(u)}, \tag{A.4}$$

$$cs(u) = \frac{cn(u)}{sn(u)},\quad ds(u) = \frac{dn(u)}{sn(u)},\quad dc(u) = \frac{dn(u)}{cn(u)}.$$

Equations for the set of functions $ns(u),\ cs(u),\ ds(u)$ are

$$\frac{d}{du}ns(u,k) = -cs(u,k)ds(u,k),$$

$$\frac{d}{du}cs(u,k) = -cs(u,k)ds(u,k),$$
$$\frac{d}{du}ds(u,k) = -ns(u,k)cs(u,k). \tag{A.5}$$

The algebraic relations between these functions are given by

$$ns^2(u,k) - cs^2(u,k) = 1, \quad ds^2(u,k) - cs^2(u,k) = 1 - k. \tag{A.6}$$

The Jacobi's imaginary transformation is given by the following mapping

$$isc(u,k') = sn(iu,k), \ cn(iu,k) = nc(u,k'), \ dn(iu,k) = dc(u,k'). \tag{A.7}$$

This set of equations obey the following equations

$$\frac{d}{du}sc(u) = nc(u)dc(u), \ \frac{d}{du}nc(u) = sc(u)dc(u), \ \frac{d}{du}dc(u) = k' \ nc(u)sc(u). \tag{A.8}$$

The algebraic relations between these functions are given by

$$nc^2(u) - sc^2(u) = 1, \quad dc^2(u) - k'sc^2(u) = 1. \tag{A.9}$$

When $k' = 0$ the set of functions $sc(u)$, $nc(u)$, $dc(u)$ are reduced to the set of hyperbolic sine-cosine functions:

$$sc(u,0) = \sinh(u), \ nc(u,0) = \cosh(u), \ dc(u,0) = 1.$$

When $k' = 1$, $nc(u,1) = dn(u,1)$. In this limit

$$sc(u,1) = tan(u), \ nc(u,1) = \frac{1}{\cos(u)}, \ dc(u,1) = \frac{1}{\cos(u)}.$$

Addition formulae for Jacobian elliptic functions

Addition formulae for the set $sn(u,k), cn(u,k), dn(u,k)$:

$$sn(u+v) = \frac{sn(u)cn(v)dn(v) + sn(v)cn(u)dn(u)}{1 - k \ sn^2(u)sn^2(v)},$$
$$cn(u+v) = \frac{cn(u)cn(v) - sn(u)sn(v)dn(u)dn(v)}{1 - k \ sn^2(u)sn^2(v)}, \tag{A.10a}$$
$$dn(u+v) = \frac{dn(u)dn(v) - k \ sn(u)sn(v)cn(u)cn(v)}{1 - k \ sn^2(u)sn^2(v)}.$$

Addition formulae for the set $sc(u,k), nc(u,k), dc(u,k)$:

$$sc(u+v) = \frac{sc(u)nc(v)dc(v) + sc(v)nc(u)dc(u)}{1 - k \ sc^2(u)sc^2(v)}, \tag{A.10b}$$
$$nc(u+v) = \frac{nc(u)nc(v) + sc(u)sc(v)dc(u)dc(v)}{1 - k \ sc^2(u)sc^2(v)},$$
$$dc(u+v) = \frac{dc(u)dc(v) + k \ sc(u)sc(v)nc(u)nc(v)}{1 - k \ sc^2(u)sc^2(v)}.$$

Bibliography

[1] Th. Kaluza, *Sitzungsberg. d. Berlin Acad. Klasse* 966 (1921).

[2] L. O' Raifertaigh and N. Strumann, *Re. Mod. Phys.* **72** 1 (2000).

[3] Yu.B. Rumer, "Function of action as a coordinate of the space" *J. Exp. Theor. Phys. (Russian)* **19** 869 (1949).

[4] H.A.Lorentz 1895 *Collected Papers* **V.I.**, Leiden

[5] H. Poincaré, *Comptes Rendues* **140** (1905) 1504

[6] A. Einstein, *Ann. d. Phys.* **b 17** (1905) 891

[7] M. Planck, *Verhandl. Deutsch. Phys. Ges.* **4** (1906), 136

[8] H. Minkowski, *Phys. Zs.* **b 10** (1909) 104

[9] R.M. Yamaleev and A.L. Fernandez Osorio, *Found. Phys. Lett.* **14** No. 4, (2001) 323

[10] R.M. Yamaleev, *Ann. of Phys. (N.Y.)* **285** **2** (2000) 1-20

[11] R.M. Yamaleev, *Ann. of Phys.(N.Y.)* **292** **2** (2001) 157

[12] M.I. Rosen 1981 "Abel's theorem on the lemniscate" *Amer. Math. Monthly* **88** 387-395

[13] N.I. Akhiezer 1970 *Elements of the theory of the elliptic functions* M., Nauka.

[14] Gudermann 1838 *Journal fur Math.* XVIII pp.12-20.
Glaisher 1882 *Messenger of Mathematics* XI p.86.

[15] V.G. Kadyshevsky, R.M. Mir-Kasimov, N.B Skachkov, *Nuovo Cimento* **A 55** (1968) 233.
R.M. Mir-Kasimov, *Physics of Particles and Nuclei* Dubna, Russia, **31 (1)** (2000) 44.

In: Relativity, Gravitation, Cosmology
Editor: V. Dvoeglazov and A. Espinoza Garrido, pp. 19-36
ISBN 1-59033-981-9

Chapter 2

The Schwarzschild Solution in a Kaluza-Klein Theory with Two Times

J. Kociński and M. Wierzbicki
Faculty of Physics, Warsaw University of Technology
Koszykowa 75, 00-662 Warszawa, Poland

Abstract

A new spherically-symmetric solution is determined in a noncompactified Kaluza-Klein theory in which a time character is ascribed to the fifth coordinate. This solution contains two independent parameters which are related with mass and electric charge. The solution exhibits a Schwarzschild radius and represents a generalization of the Schwarzschild solution in four dimensions. The parameter of the solution connected with the electric charge depends on the derivative of the fifth (second time) coordinate with respect to the ordinary time coordinate. It is shown that the perihelic motion in four-dimensional relativity has a counterpart in five dimensions in the perinucleic motion of a negatively-charged particle. If the quantization conditions of the older quantum theory are applied to that motion, an analogue of the fine-structure formula of atomic spectra is obtained.

2.1 Introduction

The notion of a second time variable is not alien to physics. In the papers of Horwitz and his coworkers [1, 2, 3, 4], a five-dimensional theory of electromagnetism was presented. In that theory the system develops on the four-dimensional spacetime manifold $(\vec{x}, t)$, according to an evolution parameter, "universal time" τ. The signatures $(4, 1)$ and $(3, 2)$ of the five-dimensional metric are considered, and the reasons for preference of the metric with signature $(3, 2)$ are explained. The five-dimensional field equations derived in [2] are called "pre-Maxwell" equations, and the respective fields which obey those equations are called "pre-Maxwell" fields. The "pre-Maxwell" equations, when referred to the Cartesian system of axes, formally represent a counterpart of Maxwell-Nordström equations (see Section 4 in [2] and [5]), with imaginary fifth coordinate, although Nordström's paper is not quoted in [2]. A possible physical meaning of the evolution parameter τ was not discussed in [2, 3, 4].

In general relativity a second time variable was introduced and discussed in the quality of a universal parametric "historical time" by Horwitz and Piron [1]. That idea was developed by Burakovsky and Horwitz [6] in their study of a five-dimensional cosmological model of Kaluza-Klein type.

In the Kaluza-Klein theory three objections were formulated against the timelike signature of the fifth coordinate [7, 8, 9]: (1) When in the five-dimensional action the integration over the fifth coordinate is performed, provided that all derivatives with respect to the fifth coordinate are omitted and the cylinder condition is accepted, the Maxwell action comes out with an opposite sign to that of the Einstein action. This is considered to be incorrect. (2) The existence of tachyons follows from the accepted cylinder condition. (3) There would appear closed time curves. The problem of closed time curves in four-dimensional gravity was investigated in a series of papers of Friedman, Thorne and their coworkers [10, 11, 12]. A particular attention was paid to the question whether closed time curves violate the causality principle. The available answer does not seem to be conclusive in that respect. In the case of the five-dimensional gravity with two time coordinates, an analogous investigation has not been undertaken. It is an open question whether the second objection concerning the existence of tachyons is relevant for noncompactified Kaluza-Klein theories. As to the first objection, it seems that the relative sign of Einstein and Maxwell actions should be the outcome of a five- or higher-dimensional relativity theory, and that an answer to the question which relative sign is correct is not yet finally settled. The attitude towards a timelike signature of the fifth coordinate is less restrained in the book of Wesson [13], in which a spacelike or a timelike signature of the fifth coordinate is admitted, depending on the physical problem in question.

The idea of two times in a physical theory is a leit-motiv in the investigations of I. Bars and his coworkers [14], [15], [16]. They have shown that two-time physics provides a new perspective for the understanding of the one-time dynamics, from a higher-dimensional point of view; from a single action formula of two-time physics, with the application of gauge theory, diverse one-time dynamical systems can be obtained.

We observe that the second time variable was employed in [5] in a discussion of the consequences following from Maxwell-Nordström equations with two time variables. These equations then describe electromagnetic phenomena together with hypothetical gravitomagnetic phenomena, in particular, gravitomagnetic waves as a counterpart of electromagnetic waves. The main results of this paper were presented in [17].

In Section 2 we touch upon the controversy concerning the relative sign of

Einstein and Maxwell actions.

In Section 3 we start from the line element which formally is identical with that of Chodos and Detweiler [18]. The difference consists in the spatial character of the fifth coordinate x^5 in [18] and the time character ascribed to this coordinate in the present case. We determine a static five-dimensional spherically-symmetric solution on the basis of the line element in which there are two time coordinates. This solution exhibits a Schwarzschild radius and represents a generalization of the four-dimensional Schwarzschild solution. The solution depends on four parameters of which two are independent and can be related with gravitational mass and electric charge. This is accomplished in Section 3 where we discuss the geodesic equation in a nearly flat space. From the geodesic equation there follows a linear relation between the ordinary time t and the second time variable u.

In Section 4 we solve the problem of the perinucleic motion of a negatively charged test particle (electron) moving in the field of the central positively charged mass. Including into the five-dimensional geometry the quantization conditions of the older quantum theory [19] we derive an analogue of Sommerfeld's relativistic energy-level formula [19]. The adiabatic invariance principle of Ehrenfest with which Bohr-Sommerfeld quantization rules are related, already played an important role in the determination of the mass spectrum of black holes [20, 21, 22, 23]. The present application of Bohr-Sommerfeld quantization rules yields another example of the role of Ehrenfest's principle in bridging the gap between general relativity and quantum theory.

2.2 The five-dimensional line element in the Kaluza-Klein theory

In the Kaluza-Klein theory the five-dimensional line element is split into the four-dimensional part and the term depending on the fifth dimension. Referring to [24] we write the line element in the five-dimensional Riemann space

$$dS^2 = \gamma_{ab}\, dx^a dx^b \quad a, b = 0, 1, 2, 3, 4 \tag{2.1}$$

where the indices $a, b = 0, 1, 2, 3$ refer to four dimensions, the label 0 refers to the time dimension, and the label 4 refers to the fifth dimension. Spacelike or timelike character of the fifth coordinate is admitted. The signature of the metric tensor therefore is $(+, -, -, -, \epsilon)$ with $\epsilon = +1$ for timelike and $\epsilon = -1$ for spacelike fifth coordinate.

A new metric tensor is defined by

$$g_{ab} = \gamma_{ab} - \frac{\gamma_{4a}\gamma_{4b}}{\gamma_{44}} \tag{2.2}$$

with the property

$$g_{44} = g_{4a} = 0 \tag{2.3}$$

We refer to [24] for this transformation, while an analogous transformation for the case of four dimensions appears with Jordan [25].

The original metric tensor can now be rewritten in the form

$$\gamma_{ab} = g_{ab} + \frac{\gamma_{4a}\gamma_{4b}}{\gamma_{44}} \tag{2.4}$$

Since $g_{ab} \neq 0$ only for $a, b = 0, 1, 2, 3$ the line element in Eq. (2.1) takes the form

$$dS^2 = g_{\mu\nu}\, dx^\mu dx^\nu + \gamma_{44}\Big(dx^4 + \frac{\gamma_{4\alpha}}{\gamma_{44}}\, dx^\alpha\Big)^2 \qquad \mu, \nu, \alpha = 0, 1, 2, 3 \tag{2.5}$$

We next define the electromagnetic four-potential

$$A_\mu = \frac{\gamma_{\mu 4}}{\gamma_{44}} \tag{2.6}$$

and the Brans-Dicke scalar field

$$\Phi^2 = \epsilon\, \gamma_{44} \tag{2.7}$$

where $\epsilon = +1$ for timelike signature and $\epsilon = -1$ for spacelike signature. The line element now takes the form

$$dS^2 = ds^2 + \epsilon\, \Phi^2 \big(dx^4 + A_\mu dx^\mu\big)^2 \tag{2.8}$$

where

$$ds^2 = g_{\mu\nu}\, dx^\mu dx^\nu\,, \qquad \mu, \nu = 0, 1, 2, 3 \tag{2.9}$$

A calculation of the components of the Riemann tensor R_{abcd} in five dimensions on the basis of the metric defined in Eq. (2.8) is largely facilitated when the calculus of exterior forms is applied. This was done by Thiry [26, 27]. The five dimensional line element in [26] is written in the form

$$d\sigma^2 = \gamma_{\mu\nu}\, dx^\mu dx^\nu = \sum_{\alpha=0}^{4} (\sigma_\alpha)^2 \tag{2.10}$$

where $\mu, \nu = 0, 1, 2, 3, 4$ and the index 0 labels the fifth dimension. Thiry introduces the new metric g_{ab} in Eq. (2.2) in which index 4 is replaced by index 0. In this metric the line element is written in the form

$$d\sigma^2 = g_{ij} dx^i dx^j + V^2 (dx^0 + \beta \varphi_i dx^i)^2\,, \quad i, j = 1, 2, 3, 4 \tag{2.11}$$

where

$$\varphi_i = \frac{\gamma_{i0}}{\beta V^2}\,, \qquad V^2 = \gamma_{00} \tag{2.12}$$

These quantities correspond to those in Eq. (2.6) and (2.7), where β is a constant. This expression for the line element is the expression in Eq. (2.8) in the case when $\epsilon = +1$, which means timelike fifth dimension. In [26] Thiry directly refers to the Kaluza-Klein theory, hence the spatial character of the fifth coordinate is implicit. In [27] the spatial character of the fifth coordinate in [26] is explicitly asserted. On the basis of the preceding argument we conclude that the spatial character assigned to the fifth coordinate in [26] and [27] is incompatible with the expression for the line element in Eq. (2.11) which appears in [26].

2.3 A spherically-symmetric solution for a two-time line element

We consider the line element in Eq. (2.1), in which, however, we change the meaning of the indices a, b. We define $a, b = 1, \ldots, 5$, where 1, 2, 3 label the space coordinates, 4 labels the ordinary time coordinate, and 5 labels the second-time coordinate.

In a flat space we introduce the Cartesian coordinates $x^1, x^2, x^3 = x, y, z$, $x^4 = ct$, $x^5 = cu$ where c denotes the speed of light in the vacuum while t and u are expressed in the units of time. The non-zero components of the metric tensor are: $\gamma_{11} = \gamma_{22} = \gamma_{33} = -1$, $\gamma_{44} = 1$ and $\gamma_{55} = 1$ since time character is ascribed to the fifth coordinate.

With

$$(x^1, x^2, x^3, x^4, x^5) = (r, \theta, \varphi, ct, cu) \tag{2.13}$$

the spherically-symmetric line element in a flat space has the form

$$dS^2 = -dr^2 - r^2(d\theta^2 + \sin^2\theta\, d\varphi^2) + c^2dt^2 + c^2du^2 \tag{2.14}$$

A general form of the spherically symmetric line element in a curved space is the following:

$$\begin{aligned} dS^2 = A(r,t,u)\, c^2dt^2 + B(r,t,u)\, dr^2 + C(r,t,u)\, cdrdt + \\ +D(r,t,u)\, (d\theta^2 + \sin^2\theta\, d\varphi^2) + E(r,t,u)\, c^2du^2 \\ +F(r,t,u)\, cdrdu + G(r,t,u)\, c^2dudt \end{aligned} \tag{2.15}$$

In an appropriate coordinate system r', t', u' we can assume that

$$C(r', t', u') = F(r', t', u') = 0 \tag{2.16}$$

and

$$D(r', t', u') = -r'^2 \tag{2.17}$$

In the following we shall omit the "prime" of the new coordinates r', t' and u' and we assume that the functions A, B, E, and G in Eq. (4) are of the form:

$$
\begin{aligned}
A(r,t,u) &= e^{\nu(r)} \qquad & B(r,t,u) &= -e^{\lambda(r)} \\
E(r,t,u) &= e^{\mu(r)} \qquad & G(r,t,u) &= \sigma(r)
\end{aligned}
\tag{2.18}
$$

with $\mu, \nu, \lambda, \sigma \to 0$ when $r \to \infty$. The line element in Eq. (6.2) now takes the form:

$$
dS^2 = -e^{\lambda} dr^2 - r^2(d\theta^2 + \sin^2\theta\, d\varphi^2) + e^{\nu} c^2 dt^2 + e^{\mu} c^2 du^2 + \sigma c^2 du dt \tag{2.19}
$$

With the speed of light in the vacuum c absorbed by the variables t and u, the respective metric tensor components are

$$
\begin{aligned}
\gamma_{11} &= -e^{\lambda}\ , \quad & \gamma_{22} &= -r^2\ , \quad & \gamma_{33} &= -r^2 \sin^2\theta\ , \\
\gamma_{44} &= e^{\nu}\ , \quad & \gamma_{55} &= e^{\mu}\ , \quad & \gamma_{45} &= \gamma_{54} = \sigma
\end{aligned}
\tag{2.20}
$$

The determinant γ of this metric tensor is given by

$$
\gamma = r^4\, e^{\lambda}\, \sin^2\theta \left(\sigma^2 - e^{\mu+\nu}\right) \tag{2.21}
$$

The line element in Eq. (2.19) is analogous to that considered by Chodos and Detweiler in Eq. (17) of [18], however, in that paper the fifth coordinate is a space coordinate, while we assign a time character to the fifth coordinate. Consequently, the spherically symmetric solution connected with the line element in Eq. (2.19) will be of a different form that in [18].

When all derivatives with respect to the times t and u are omitted, denoting by a "prime", the derivative d/dr we obtain the following non-zero components of the contracted Riemann tensor

$$
\begin{aligned}
R_{11} = \frac{1}{4\,r\,(\sigma^2 - e^{\mu+\nu})^2} \Big\{ & 2\,r\,\left(-\sigma^2 - e^{\mu+\nu}\right) \sigma'^{\,2} \\
& - \left(\sigma^2 - e^{\mu+\nu}\right) \lambda' \left(4\,\sigma^2 - 4\,e^{\mu+\nu} - e^{\mu+\nu}\, r\, \mu' - e^{\mu+\nu}\, r\, \nu'\right) \\
& -2\,\sigma\, r\, \sigma' \left[\left(\sigma^2 - e^{\mu+\nu}\right) \lambda' - 2\,e^{\mu+\nu}\,(\mu' + \nu')\right] \\
& + r \Big[e^{\mu+\nu}(e^{\mu+\nu} - 2\sigma^2)(\mu'^2 + \nu'^2) - 2\,e^{\mu+\nu}\,\sigma^2\,\mu'\,\nu' \\
& + 4\,\sigma^3\,\sigma'' - 4\,e^{\mu+\nu}\,\sigma\,\sigma'' + 2e^{\mu+\nu}(e^{\mu+\nu} - \sigma^2)(\mu'' + \nu'') \Big] \Big\}
\end{aligned}
\tag{2.22}
$$

$$
\begin{aligned}
R_{22} = -\frac{e^{-\lambda}}{2\,(\sigma^2 - 2e^{\mu+\nu})} \Big[& -2\,\sigma\, r\, \sigma' + \sigma^2 \left(-2 + 2\,e^{\lambda} + r\,\lambda'\right) \\
& - e^{\mu+\nu} \left(-2 + 2\,e^{\lambda} + r\,\lambda' - r\,\mu' - r\,\nu'\right) \Big]
\end{aligned}
\tag{2.23}
$$

$$R_{33} = R_{22} \sin^2\theta \tag{2.24}$$

$$\begin{aligned} R_{44} = & \frac{-e^{-\lambda+\nu}}{4\,r\,(\sigma^2 - e^{\mu+\nu})} \Big[2\,r\,\sigma'^2 - 2\,\sigma\,r\,\sigma'\,\nu' \\ & + \left[4\,\sigma^2 - 4\,e^{\mu+\nu} - \sigma^2\,r\,\lambda' + e^{\mu+\nu}\,r\,(\lambda' - \mu')\right]\,\nu' \\ & + r\,\left(2\,\sigma^2 - e^{\mu+\nu}\right)\,\nu'^2 + 2\,r\,\left(\sigma^2 - e^{\mu+\nu}\right)\,\nu'' \Big] \end{aligned} \tag{2.25}$$

$$\begin{aligned} R_{45} = R_{54} = & \frac{1}{4\,e^{\lambda}\,r\,(\sigma^2 - e^{\mu+\nu})} \Big[\sigma^2\,\sigma'\,(-4 + r\,\lambda') \\ & - e^{\mu+\nu}\,\sigma'\,(-4 + r\,\lambda' + r\,\mu' + r\,\nu') \\ & + 2\,r\,\left(e^{\mu+\nu}\,\sigma\,\mu'\,\nu' - \sigma^2\,\sigma'' + e^{\mu+\nu}\,\sigma''\right) \Big] \end{aligned} \tag{2.26}$$

$$\begin{aligned} R_{55} = & \frac{-e^{-\lambda+\mu}}{4\,r\,(\sigma^2 - e^{\mu+\nu})} \Big[2\,r\,\sigma'^2 - 2\,\sigma\,r\,\sigma'\,\mu' + r\,\left(2\,\sigma^2 - e^{\mu+\nu}\right)\,\mu'^2 \\ & + \mu'\,\left(4\,\sigma^2 - 4\,e^{\mu+\nu} - \sigma^2\,r\,\lambda' + e^{\mu+\nu}\,r\,\lambda' - e^{\mu+\nu}\,r\,\nu'\right) \\ & + 2\,r\,\left(\sigma^2 - e^{\mu+\nu}\right)\,\mu'' \Big] \end{aligned} \tag{2.27}$$

These are equated to zero, yielding five equations for the four unknown functions $\mu, \nu, \lambda, \sigma$ in Eq. (6.5).

It can be verified that a solution of the equations

$$R_{11} = R_{22} = R_{44} = R_{55} = R_{45} = 0 \tag{2.28}$$

is given by the following functions:

$$e^{\nu} = 1 - \frac{\mathcal{G}}{r}, \quad e^{\mu} = 1 - \frac{\mathcal{C}}{r}, \quad e^{\lambda} = \left(1 - \frac{\mathcal{R}}{r}\right)^{-1}, \quad \sigma = \frac{\mathcal{P}}{r} \tag{2.29}$$

where the real parameters $\mathcal{G}, \mathcal{C}, \mathcal{R}$ and $\mathcal{P}$ satisfy the conditions

$$\mathcal{R} = \mathcal{G} + \mathcal{C}, \quad \mathcal{P}^2 = \mathcal{G}\mathcal{C} \tag{2.30}$$

The parameters $\mathcal{G}$ and $\mathcal{C}$ will be related with gravitational mass and electric charge, respectively, on the basis of the linearized form of the geodesic equation.

2.4 The parameters in the spherically-symmetric solution

We consider the two-time-independent metric tensor of the form:

$$\gamma_{ab} = \gamma^{(5)}_{ab} + \eta_{ab}\,, \quad a, b = 1, \ldots, 5 \tag{2.31}$$

where $\gamma^{(5)}_{ab}$ is the five-dimensional flat-space metric tensor specified as at the beginning of Section 3, and η_{ab} represents a small perturbation, due to the presence of a gravitating body with an electric charge. The perturbation vanishes very far from the body. To show that the η_{ab} terms are the agents of gravitational and electrostatic forces we consider the geodesic equation of motion in the Riemann space with the above metric. We assume that the velocity of a test particle (with mass and electric charge) along the geodesic line is much smaller than the speed of light c. Using the nearly flat-space metric tensor in Eq. (6.11) we then obtain the line element

$$dS^2 = -(dx^1)^2 - (dx^2)^2 - (dx^3)^2 + (dx^4)^2 + (dx^5)^2 + \eta_{ab}dx^a dx^b \tag{2.32}$$

from which we obtain:

$$\left(\frac{dS}{dt}\right)^2 = c^2\left[1 - \frac{v^2}{c^2} + \left(\frac{du}{dt}\right)^2 + \frac{1}{c^2}\eta_{ab}\frac{dx^a}{dt}\frac{dx^b}{dt}\right] \tag{2.33}$$

where $v^j = dx^j/dt$, $j = 1, 2, 3$. Retaining in Eq. (6.13) the terms of the first order in v/c we obtain:

$$\left(\frac{dS}{dt}\right)^2 \approx c^2\left[1 + \left(\frac{du}{dt}\right)^2(1 + \eta_{55}) + \eta_{44} + 2\eta_{45}\frac{du}{dt}\right] \tag{2.34}$$

We had to assume that $v^2/c^2 \ll (du/dt)^2$, since otherwise we would have to omit $(du/dt)^2$ together with v^2/c^2.

We next apply the same approximations to the geodesic equation

$$\frac{d^2x^a}{dS^2} + \Gamma^a_{bc}\frac{dx^b}{dS}\frac{dx^c}{dS} = 0 \tag{2.35}$$

From the form of the metric in Eq. (6.12) follows that each Christoffel symbol linearly depends on the perturbation η_{ab}. With the accuracy to terms of order v/c we obtain the equality

$$\Gamma^a_{bc}\frac{dx^b}{dS}\frac{dx^c}{dS} = c^2\left[\Gamma^a_{44} + 2\frac{du}{dt}\Gamma^a_{45} + \left(\frac{du}{dt}\right)^2\Gamma^a_{55}\right]\left(\frac{dt}{dS}\right)^2 \tag{2.36}$$

From Eq. (6.15) and (6.17) follows the equation

$$\frac{d^2x^a}{dt^2} = c^2\left[\Gamma^a_{44} + 2\frac{du}{dt}\Gamma^a_{45} + \left(\frac{du}{dt}\right)^2\Gamma^a_{55}\right] \tag{2.37}$$

Since η_{ab} are independent of t and u the Christoffel symbols in Eq. (6.19) vanish for $a = 4, 5$. For $a = 1, 2, 3$ we have

$$\Gamma^a_{44} = \frac{1}{2}\frac{\partial \eta_{44}}{\partial x^a}, \quad \Gamma^a_{45} = \frac{1}{2}\frac{\partial \eta_{45}}{\partial x^a}, \quad \Gamma^a_{55} = \frac{-1}{2}\frac{\partial \eta_{55}}{\partial x^a} \tag{2.38}$$

and from Eq. (6.19), for $a = 5$ we find that $d^2u/dt^2 = 0$, hence

$$u = wt + u_0 \tag{2.39}$$

where w and u_0 are constants. With $u_0 = 0$, $w > 0$, considering Eqs. (6.20), and with $w = du/dt$ we obtain from Eq. (6.19) the equation

$$\frac{d^2x^a}{dt^2} = -\frac{c^2}{2}\left[\frac{\partial \eta_{44}}{\partial x^a} + 2w\frac{\partial \eta_{45}}{\partial x^a} + w^2\frac{\partial \eta_{55}}{\partial x^a}\right] \tag{2.40}$$

On the basis of this equation we will determine the parameters $\mathcal{G}$ and $\mathcal{C}$ in Eqs. (6.9) and (6.10).

We assume that the test particle is an electron with mass m. Multiplying both sides of Eq. (6.22) with m, we can identify the first term on the r.h.s. with the gravitational force acting on the electron mass m, and to one of the two remaining terms we can ascribe the meaning of the Coulomb force acting on the electron charge e. It will appear that we have to relate the third term on the r.h.s. of Eq. (6.22) with the Coulomb force.

We begin with the first term on the r.h.s. of Eq. (6.22) and write:

$$\left(m\frac{d^2x^a}{dt^2}\right)_{\text{mechanical}} = -m\frac{c^2}{2}\frac{\partial \eta_{44}}{\partial x^a} = -m\frac{\partial \psi}{\partial x^a} \tag{2.41}$$

with ψ denoting the gravitational potential of the mass M, where κ is the gravitational constant,

$$\psi = -\kappa\frac{M}{r} \tag{2.42}$$

thus obtaining the equality

$$\eta_{44} = \frac{2}{c^2}\psi \tag{2.43}$$

and from this and from Eq. (6.12), the equality

$$\gamma_{44} = 1 + \eta_{44} \tag{2.44}$$

From the first of Eqs. (6.9) and from Eqs. (6.23), (2.43) and (2.44) we then find that

$$\mathcal{G} = \frac{2\kappa M}{c^2} \tag{2.45}$$

as in the case of the Schwarzschild solution in 4 dimensions.

We next consider the third term on the r.h.s of Eq. (6.22). Let M denote the proton mass and let φ denote the electrostatic potential of the proton charge Q

$$\varphi = \frac{Q}{4\pi\varepsilon_0 r} \tag{2.46}$$

where ϵ_0 denotes the vacuum electric permeability. On the basis of Eq. (6.22) we write

$$\left(m\frac{d^2x^a}{dt^2}\right)_{\text{electrical}} = -\frac{1}{2}mc^2w^2\frac{\partial\eta_{55}}{\partial x^a} = -e\frac{\partial\varphi}{\partial x^a} \tag{2.47}$$

for the electrostatic force acting on the electric charge e connected with the mass m. From Eq. (2.47) and (2.48) we obtain

$$\frac{1}{2}c^2w^2\eta_{55} = \frac{e}{m}\varphi = \frac{Qe}{4\pi\varepsilon_0 rm} \tag{2.48}$$

and hence

$$\eta_{55} = \frac{2eQ}{4\pi\varepsilon_0 mc^2w^2r} \tag{2.49}$$

On the other hand from the second of Eqs. (6.9) and from Eq. (6.12) we obtain

$$\gamma_{55} = 1 + \eta_{55} = 1 - \frac{\mathcal{C}}{r} \tag{2.50}$$

From Eqs. (2.49) and (2.50) we find that

$$\mathcal{C} = \frac{-2eQ}{4\pi\varepsilon_0 mc^2w^2} \tag{2.51}$$

From the second of Eqs. (6.10) and from Eq. (2.46) follows that $\mathcal{C}$ must be positive. This implies that when the charge Q is assumed positive, the charge e must be negative and vice versa. When e and Q are of the same sign the second condition in Eq. (6.10) cannot be fulfilled. This formula depends on the parameter $w^2 = (du/dt)^2$, which in connection with Eq. (6.14) has to be much larger than v^2/c^2.

We now can determine the parameter $\mathcal{P}$ in Eq. 6.10). Owing to Eq. (2.51) we obtain

$$\mathcal{P} = \sqrt{\mathcal{GC}} = \frac{1}{wc^2}\sqrt{\frac{\kappa M|eQ|}{\pi\varepsilon_0 m}} \tag{2.52}$$

where $-eQ$ in Eq. (2.51) has been replaced by $|eQ|$, since e and Q must have opposite signs. With the gravitational constant $\kappa = 6.673 \times 10^{-11}\,\mathrm{N}\cdot\mathrm{M}^2\cdot\mathrm{K}$, the vacuum electric permeability $\varepsilon_0 = 0.885 \times 10^{-11}\,\mathrm{Q}\cdot\mathrm{V}^{-1}\cdot\mathrm{M}^{-1}$, with $e/m = 1.76 \times 10^{11}\,\mathrm{Q}\cdot\mathrm{K}$, and the absolute value of electron charge $|e| = 1.6 \times 10^{-19}\,\mathrm{Q}$, where N=newton, M=meter, K=kilogram, Q=coulomb, V=volt [28] identifying the mass M in Eq. (2.45) with the proton mass we find from Eq. (2.45) that

$$\mathcal{G} \approx 2.4 \times 10^{-54}\,\mathrm{M} \tag{2.53}$$

With Q denoting the proton charge, from Eq. (2.51) we find that

$$\mathcal{C} \approx 5.6\, w^{-2} \times 10^{-15}\,\mathrm{M} \tag{2.54}$$

and then with $Q = |e|$ from Eq. (2.52) we obtain

$$\mathcal{P} \approx 1.2\, w^{-1} \times 10^{-34}\,\mathrm{M} \tag{2.55}$$

We now can answer the question for the Schwarzschild radius. From Eqs. (6.10), (2.45) and (2.51) we find that in Eq. (19)

$$\mathcal{R} = \mathcal{G} + \mathcal{C} = 2.4 \times 10^{-54}\,\mathrm{M} + 5.6\, w^{-2} \times 10^{-15}\,\mathrm{M} \tag{2.56}$$

If w^{-2} is not extremely small, the Schwarzschild radius is determined by the parameter $\mathcal{C}$ connected with the electric charge of the proton.

We now explain why it is not possible to relate the term $\mathcal{P}/r$ with the Coulomb potential. If $\mathcal{P}/r$ were related with the Coulomb potential, from Eq. (6.22) we then would obtain:

$$\left(m\frac{d^2x^a}{dt^2}\right)_{\text{electrical}} = -mc^2 w \frac{\partial \eta_{45}}{\partial x^a} = -e\frac{\partial \varphi}{\partial x^a} \tag{2.57}$$

with φ defined in Eq. (2.46). From Eqs. (2.46) and (2.57) we would obtain:

$$\eta_{45} = \frac{eQ}{4\pi\varepsilon_0 r m c^2 w} \tag{2.58}$$

and since from Eqs. (2.19) and (6.12) we have

$$\gamma_{45} = \eta_{45} = \frac{\mathcal{P}}{r} \tag{2.59}$$

we would find

$$\mathcal{P} = \frac{eQ}{4\pi\varepsilon_0 m c^2 w} \tag{2.60}$$

The parameter $\mathcal{C}$ now is obtained from Eq. (6.10)

$$\mathcal{C} = \mathcal{P}^2 \mathcal{G}^{-1} \tag{2.61}$$

The parameter $\mathcal{R}$ is determined in Eq. (6.10). Inserting into Eq. (2.61) the numerical values for $\mathcal{P}$ and $\mathcal{G}$ given in Eqs. (2.55) and (2.53) we find that $\mathcal{C} \approx 3\, w^{-2} \times 10^{24}\,\mathrm{M}$, hence, unless w^{-2} is extremely small, the respective $\mathcal{R} = \mathcal{G} + \mathcal{C}$ is unacceptable as a candidate for the Schwarzschild radius of the proton. This seems to indicate that we cannot identify the force connected with η_{45} with the Coulomb force between the central charge Q and the charge e of a test particle.

2.5 The relativistic energy-level formula

We set up the variational problem for the square of the interval in Eq. (2.19) (see for example [29]) in the form

$$\delta\int\Big[\Big(1-\frac{\mathcal{G}}{r}\Big)c^2\,\dot{t}^2+\Big(1-\frac{\mathcal{C}}{r}\Big)c^2\,\dot{u}^2+\frac{\sqrt{\mathcal{GC}}}{r}c^2\,\dot{t}\dot{u}-\Big(1-\frac{\mathcal{G}+\mathcal{C}}{r}\Big)^{-1}\dot{r}^2-r^2(\dot{\theta}^2+\dot{\varphi}^2\,\sin^2\theta)\Big]dS=0 \tag{2.62}$$

where the "dot" denotes d/dS. The Euler-Lagrange equations for θ, φ, t and u yield the equalities:

$$\frac{d}{dS}(r^2\,\dot{\theta})-r^2\dot{\varphi}^2\sin\theta\cos\theta=0 \tag{2.63}$$

$$\frac{d}{dS}(r^2\dot{\varphi}\sin^2\theta)=0 \tag{2.64}$$

$$\frac{d}{dS}\Big[2\Big(1-\frac{\mathcal{G}}{r}\Big)\dot{t}+\frac{\sqrt{\mathcal{GC}}}{r}\dot{u}\Big]=0 \tag{2.65}$$

$$\frac{d}{dS}\Big[2\Big(1-\frac{\mathcal{C}}{r}\Big)\dot{u}+\frac{\sqrt{\mathcal{GC}}}{r}\dot{t}\Big]=0 \tag{2.66}$$

Dividing the expression for the interval in Eq. (2.19) by dS^2 we obtain the equation for $\dot{r}$

$$1=\Big(1-\frac{\mathcal{G}}{r}\Big)c^2\dot{t}^2+\Big(1-\frac{\mathcal{C}}{r}\Big)c^2\dot{u}^2+\frac{\sqrt{\mathcal{GC}}}{r}c^2\dot{t}\dot{u}-\Big(1-\frac{\mathcal{G}+\mathcal{C}}{r}\Big)^{-1}\dot{r}^2-r^2(\dot{\theta}^2+\dot{\varphi}^2\sin^2\theta) \tag{2.67}$$

We assign the charge $Z|e|$ to the central mass and will determine the perinucleic motion of an electron in the Coulomb field. We observe that the term with $\sqrt{\mathcal{GC}}$ determines the main influence of the gravitational field, connected with the central singularity, on the energy levels. This influence is very small in comparison with the influence of the Coulomb field represented by the parameter $\mathcal{C}$. In Eqs. (2.65), (2.66), and (2.67) we omit the terms containing the factors $\mathcal{G}/r$ or $\sqrt{\mathcal{GC}}/r$, since they are small in comparison with the terms containing $\mathcal{C}/r$, thus obtaining from Eq. (2.67) the equation

$$1=c^2\dot{t}^2+\Big(1-\frac{\mathcal{C}}{r}\Big)c^2\dot{u}^2-\Big(1-\frac{\mathcal{C}}{r}\Big)^{-1}\dot{r}^2-r^2(\dot{\theta}^2+\dot{\varphi}^2\sin^2\theta) \tag{2.68}$$

By an appropriate orientation of the coordinate axes we can make $\theta=\pi/2$ and $d\theta/dS=\dot{\theta}=0$, for some initial value of S. From Eq. (2.63) then follows that for all values of S we have $\theta = \pi/2$. Substituting this value of θ into Eq. (2.64) we obtain

$$r^2 \frac{d\varphi}{dS} = k = \text{const} \tag{2.69}$$

while from Eqs. (2.65) and (2.66) we obtain

$$\frac{dt}{dS} = \tau = \text{const} \tag{2.70}$$

and

$$\frac{du}{dS}\left(1 - \frac{\mathcal{C}}{r}\right) = \rho = \text{const} \tag{2.71}$$

Substituting the expressions in Eqs. (2.69), (2.70) and (2.71) into Eq. (2.68) we obtain the equation for $r = r(S)$

$$\left(\frac{dr}{dS}\right)^2 = [c^2(\tau^2 + \rho^2) - 1] + (1 - c^2\tau^2)\frac{\mathcal{C}}{r} - \frac{k^2}{r^2} + \frac{\mathcal{C}k^2}{r^3} \tag{2.72}$$

From this equation in the new variable $v = 1/r$, in the customary way [29] we obtain the equation

$$v'' + v = \mathcal{A} + \frac{\beta}{\mathcal{A}} v^2 \tag{2.73}$$

where

$$\mathcal{A} = \frac{(1 - c^2\tau^2)\mathcal{C}}{2k^2} \qquad \text{and} \qquad \beta = \frac{3}{2}\mathcal{AC} \tag{2.74}$$

Eq. (2.73) determines the perinucleic motion.

We now intend to determine a formula for the energy levels of the test particle. From Eq. (2.72) we obtain the expression

$$\frac{dr}{dS} = \frac{dr}{dt}\frac{dt}{dS} = \tau \frac{dr}{dt} = \sqrt{[c^2(\tau^2 + \rho^2) - 1] + (1 - c^2\tau^2)\frac{\mathcal{C}}{r} - \frac{k^2}{r^2} + \frac{\mathcal{C}k^2}{r^3}} \tag{2.75}$$

We take over from Sommerfeld (p. 277 in [19]) the quantization conditions

$$\oint m \frac{dr}{dt} dr = n_r h\,, \quad n_r = 0, 1, 2, \ldots \tag{2.76}$$

$$\int_0^{2\pi} m r^2 \frac{d\varphi}{dt} d\varphi = n_\varphi h\,, \quad n_\varphi = 1, 2, 3, \ldots \tag{2.77}$$

where m denotes the electron mass, and h is Planck's constant. From Eq. (2.75) we find that

$$m\frac{dr}{dt} = \sqrt{\frac{m^2}{\tau^2}\left\{[c^2(\tau^2+\rho^2)-1] + (1-c^2\tau^2)\frac{\mathcal{C}}{r} - \frac{k^2}{r^2} + \frac{\mathcal{C}k^2}{r^3}\right\}} \tag{2.78}$$

From Eqs. (2.69) and (2.70) and from Eq. (2.77) we find that

$$k = \frac{\tau}{m} n_\varphi \hbar \tag{2.79}$$

where $\hbar = h/2\pi$. The integral in Eq. (2.76) with the integrand given in Eq. (2.78) was calculated in [19]. It has the value

$$\begin{aligned} I = \oint \sqrt{A_0 + 2\frac{A_1}{r} + \frac{A_2}{r^2} + \frac{A_3}{r^3}}\,dr = \\ -2\pi i\left(\sqrt{A_2} - \frac{A_1}{\sqrt{A_0}} - \frac{A_1 A_3}{2A_2\sqrt{A_2}}\right) \end{aligned} \tag{2.80}$$

where $\sqrt{A_2}$ is negative imaginary. Comparing Eq. (2.78) with the integrand on the l.h.s. of Eq. (2.80) we find that

$$\begin{aligned} A_0 &= \frac{m^2}{\tau^2}\left[c^2(\tau^2+\rho^2)-1\right] \qquad & A_1 &= \frac{m^2}{2\tau^2}(1-c^2\tau^2)\,\mathcal{C} \\ A_2 &= -\frac{m^2k^2}{\tau^2} = -n_\varphi^2\hbar^2 \qquad & A_3 &= \frac{m^2}{\tau^2}\mathcal{C}k^2 = \mathcal{C}n_\varphi^2\hbar^2 \end{aligned} \tag{2.81}$$

The three terms on the r.h.s. of Eq. (2.80) take the form

$$-2\pi i\sqrt{A_2} = -2\pi n_\varphi \hbar \tag{2.82}$$

$$2\pi i\frac{A_1}{\sqrt{A_0}} = 2\pi i\frac{m(1-c^2\tau^2)\,\mathcal{C}}{2\tau\sqrt{c^2(\tau^2+\rho^2)-1}} \tag{2.83}$$

$$\pi i\frac{A_1A_3}{A_2\sqrt{A_2}} = 2\pi\frac{(1-c^2\tau^2)\,\mathcal{C}^2m^2}{4n_\varphi\hbar\tau^2} \tag{2.84}$$

From Eqs. (2.75), (2.78), (2.80) and (2.82) through (2.84), introducing the expression for $\mathcal{C}$ given in Eq. (2.51), with $\alpha = e^2/4\pi\varepsilon_0\hbar c$ we obtain the equality

$$\frac{-iZ\alpha(1-c^2\tau^2)}{\tau c\sqrt{c^2(\tau^2+\rho^2)-1}}\left(\frac{dt}{du}\right)^2 = n_r + n_\varphi - \left[\frac{1-c^2\tau^2}{c^2\tau^2}\left(\frac{dt}{du}\right)^4\right]\frac{\alpha^2Z^2}{n_\varphi} \tag{2.85}$$

Writing $du/dt = w$ we find from Eqs. (6.13), (2.70) and (2.71) with $v^2/c^2 \ll w^2$, as it was assumed in Section 4, that

$$c^2\tau^2 \approx \frac{1}{1+w^2} \tag{2.86}$$

and

$$c^2\rho^2 \approx \frac{w^2}{1+w^2}\Big(1-\frac{C}{r}\Big) \tag{2.87}$$

Hence w^2 is approximately expressed through the constant of motion τ. Utilizing these equalities we can rewrite Eq. (2.85) in the form

$$1+w^2\Big(2\frac{C}{r}-\frac{C^2}{r^2}\Big) = 1+\frac{\alpha^2 Z^2}{\Big(n_r+n_\varphi-\dfrac{\alpha^2 Z^2}{w^2 n_\varphi}\Big)^2} \tag{2.88}$$

and compare it with Sommerfeld's formula for energy levels of an electron in the relativistic Kepler motion in a flat space in Eq. (26) on p. 278 of [19] which is

$$\Big(1+\frac{W}{m_0c^2}\Big) = \left[1+\frac{\alpha^2 Z^2}{\big(n_r+\sqrt{n_\varphi^2-\alpha^2 Z^2}\big)^2}\right]^{-\frac{1}{2}} \tag{2.89}$$

where $W = E - m_0c^2$ and E = total energy. The inverse square root of the l.h.s. of Eq. (2.88) can be equalized to the total energy E. On the r.h.s. of Eq. (2.89) the first two terms of a series expansion of the square root are equal to the respective two terms in Eq. (2.88), when we put $w^2 = 2$. This means that for $\alpha^2 Z^2/2n_\varphi^2 \ll 1$ and $w^2 = 2$ the r.h.s. of Sommerfeld's formula in Eq. (2.89) coincides with the inverse square root of the r.h.s. of Eq. (2.88). Eq. (2.88) therefore represents an analogue of Sommerfeld's formula for energy levels of a hydrogen-like atom for sufficiently small values of Z.

2.6 Conclusions and discussion

There exists an extensive literature concerning spherically-symmetric solutions in the Kaluza-Klein theory. We only name the five-dimensional spherically symmetric solutions determined by Chodos and Detweiler [18], by Ponce de Leon and Wesson [30] and by Wesson [13]. Those solutions are based on the assumption of a spatial character of the fifth coordinate. A thorough discussion of those solutions is given in Overduin and Wesson [9] and in Wesson [13].

Assuming that in a non-compactified Kaluza-Klein theory the fifth coordinate x^5 has time character, we have determined a Schwarzschild type solution of the five-dimensional Einstein equations in the vacuum. The two independent parameters of that solution have been related with mass and electric charge, respectively. The solution exhibits a Schwarzschild radius whose magnitude is predominantly determined by the electric-charge parameter.

The perihelic motion in four-dimensional relativity has a counterpart in the perinucleic motion of an electron in a Kaluza-Klein theory with two times. If the quantization conditions of the older quantum theory are included into the five-dimensional geometry, the perinucleic motion of an electron leads to the fine structure of line spectra, which in the limit of $Z^2\alpha^2/n_\varphi^2 \ll 1$ is analogous to that determined by Sommerfeld's formula for hydrogen-like atoms.

The parameter $\mathcal{C}$ which determines the Schwarzschild radius and the parameter $\mathcal{P}$ connected with a new force depend on the derivative du/dt of the second time coordinate u with respect to the ordinary time coordinate t. Their numerical values therefore hinge on the magnitude of that quantity. The indicated physical meaning of those parameters, however, is not impaired by the lack of knowledge of the magnitude of the quantity du/dt as long as it is not extremely large or extremely small in comparison with 1.

Acknowledgment

We thank the referees for their comments and for pointing out to us several important papers.

Bibliography

[1] Horwitz L. P. and Piron C., (1973) *Helv. Phys. Acta* **46** 316

[2] Saad D., Horwitz L. P. and Arshansky R. I., (1989) *Found. Phys.* **19** 1125

[3] Shnerb N. and Horwitz L. P. , (1993) *Phys. Rev.* A**48** 4068

[4] Land M. C., Shnerb N. and Horwitz L. P., (1995) *J. Math. Phys.* **36** 3263

[5] Kociński J., (2003) *Electromagnetic Phenomena* **T3** No. 1(9) 106

[6] Burakovsky L. and Horwitz L. P. (1995) *Gen. Rel. Grav* **27** 1043

[7] Appelquist T., Chodos A. and Freund P. G .O., (1987) *Modern Kaluza-Klein Theories*, Addison-Wesley, Reading

[8] Bailin D. and Love L., (1987) *Kaluza-Klein theories*, *Rep. Prog. Phys.* **50**, 1087

[9] Overduin J. M. and Wesson P. S., (1997) *Phys. Reports* **283**, 303

[10] Friedman J. L., M. S. Morris, I. D. Novikov, F. Echeverria, G. Klinkhammer, K.S. Thorne and U. Yurtsever, (1990) *Phys. Rev. D* **42** 1915

[11] Thorne K. S., (1991) *Annals N. Y. Acad. Sci.* **631** 182

[12] Friedman J. L., (1991) *Annals N .Y. Acad. Sci.* **647** 620

[13] Wesson P. S., (2000) *Space-Time-Matter, Modern Kaluza-Klein Theory,* World Scientific Singapore

[14] Bars I. and Kounnas C., (1997) *Phys. Rev. D* **56** 3664

[15] Bars I., Deliduman C. and Andreev O., (1998) *Phys. Rev. D* **58** 066004

[16] Bars I., Deliduman C., Minic D., (1999) *Phys. Rev. D* **59** 125004

[17] Kociński J. and Wierzbicki M., arXiv:gr-qc/0110075

[18] Chodos A. and Detweiler S., (1982) *Gen. Rel. Grav.* **14**, 879

[19] Sommerfeld A., (1960) *Atombau und Spektrallinien I. Band*, Friedr. Vieweg & Sohn, Braunschweig

[20] Bekenstein J. D., Mukhanov V. F. arXiv:gr-qc/9505012v1

[21] Bekenstein J. D., arXiv:gr-qc/9710076v2

[22] Bekenstein J. D., arXiv:gr-qc/0107049v2

[23] Barvinsky A., Das S., Kunstatter G., arXiv:hep-th/0209039v1

[24] Wesson P. S., Ponce de Leon J., (1995) *Astron. Astrophys.* **294** 1

[25] Jordan P, (1955) *Schwerkraft und Weltall*, Friedr. Vieweg & Sohn, Braunschweig

[26] Thiry Y., (1948) *Comptes rendus* **226**(3) 216

[27] Thiry Y., (1948) *Comptes rendus* **226**(23) 1881

[28] Sommerfeld A., (1963) *Electrodynamics, Lectures on Theoretical Physics, vol. III*, Academic Press, New York

[29] Adler R., M. Basin and M. Schiffer, (1965) *Introduction to General Relativity*, McGraw-Hill Book Company, New York

[30] Ponce de Leon J. and Wesson P. S., (1993) *J. Math. Phys.* **34** 4080

In: Relativity, Gravitation, Cosmology
Editor: V. Dvoeglazov and A. Espinoza Garrido, pp. 37-42
ISBN 1-59033-981-9

Chapter 3

Cartan's Torsion: Necessity and Observational Evidence

Rainer W. Kühne
Lechstr. 63, 38120 Braunschweig, Germany
kuehne70@gmx.de

Abstract

This article starts with the mathematical definition, concrete description, and physical meaning of Cartan's torsion. I proceed with the argumentation that torsion is required for the description of intrinsic spin. Moreover I argue that the duality between curvature and torsion is analogous to the duality between electricity and magnetism. I conclude this article by pointing out that the aligned rotation axes of the galaxies of the Perseus-Pisces supercluster may be interpreted as a topological defect generated by torsion.

3.1 What is Cartan's Torsion?

When a four-vector C^k is displaced parallel from the four-position x^k to x^k+dx^k, then it changes according to the prescription,

$$dC^k = -\Gamma^k_{ij}(x)C^j dx^i. \tag{3.1}$$

This is the definition for the position-dependent affine connection Γ^k_{ij}. According to general relativity [1], it has only a symmetric part,

$$\{\}^k_{ij} = \frac{1}{2}(\Gamma^k_{ij} + \Gamma^k_{ji}), \tag{3.2}$$

which is called "Christoffel symbol." The anti-symmetric part of the affine connection is called "Cartan's torsion" [2],

$$T^k_{ij} = \frac{1}{2}(\Gamma^k_{ij} - \Gamma^k_{ji}). \tag{3.3}$$

According to general relativity, the torsion tensor is zero. The introduction of a nonzero torsion tensor means therefore an extension of general relativity.

Quite remarkably, the torsion tensor transforms as a tensor under local Lorentz transformations, whereas the Christoffel symbol does not.

The torsion tensor can be viewed as the translational field strength. It represents a closure failure of infinitesimal displacements. In spacetimes which include torsion, infinitesimal parallelograms do not close.

We know from Einstein's general relativity [1] that gravitational mass is connected with curvature via

$$G^{ij} = \kappa \Sigma^{ij}, \tag{3.4}$$

where

$$G^{ij} = R^{ij} - \frac{1}{2} g^{ij} R^k_k \tag{3.5}$$

is the Einstein tensor, Σ^{ij} is the stress-energy (energy-momentum) tensor, R^{ij} is the Ricci tensor, g^{ij} is the metric tensor, R^k_k is the Ricci scalar, and $\kappa = -8\pi G/c^4$ is the Einstein constant.

Analogously, intrinsic spin is connected with Cartan's torsion via

$$T^{ijk} = \kappa \tau^{ijk}, \tag{3.6}$$

where τ^{ijk} is the spin tensor [3]. The equations (4) and (6) show the duality between mass and spin and between curvature and torsion, respectively.

Directly from the definition of the affine connection, Eq. (1), one obtains the differential equation of autoparallel curves,

$$\frac{d^2 x^k}{ds^2} + \Gamma^k_{ij} \frac{dx^i}{ds} \frac{dx^j}{ds} = 0, \tag{3.7}$$

where the infinitesimal interval ds between x^k and $x^k + dx^k$ is given by

$$ds^2 = g_{ij}(x) dx^i dx^j. \tag{3.8}$$

Quite remarkably, only the symmetric part of the metric tensor contributes to the square of the infinitesimal interval.

Readers who would like to learn more about the formalism of torsion are invited to read the excellent reviews, Ref. [4].

3.2 Why Do We Need Torsion?

The energy-momentum tensor Σ^{ij} of a Dirac field Ψ (spin 1/2 field [5]) is antisymmetric [6],

$$\Sigma^{ij} = -\frac{\hbar c}{2} [(\nabla^i \bar{\Psi}) \gamma^j \Psi - \bar{\Psi} \gamma^j \nabla^i \Psi], \tag{3.9}$$

where

$$\nabla_i = \partial_i + ieA_i \tag{3.10}$$

is the covariant derivative. By contrast, the energy-momentum tensor of general relativity [1] is symmetric. In order to couple a spinor field (Dirac field) to a gravitational field, one has to use an energy-momentum tensor which includes anti-symmetric parts. Therefore general relativity has to be generalized by the introduction of Cartan's torsion [3].

I have shown that the duality between mass and spin is analogous to the duality between electric charge and magnetic charge [7]. The electric-magnetic duality is,

$$J^i = \partial_j F^{ji} \tag{3.11}$$

$$j^i = \partial_j f^{ji}, \tag{3.12}$$

where J^i is the electric four-current, j^i is the magnetic four-current, and the field strength tensors are given by,

$$F^{ji} = \partial^j A^i - \partial^i A^j \tag{3.13}$$

$$f^{ji} = \partial^j a^i - \partial^i a^j, \tag{3.14}$$

where A^j is the electric four-potential which corresponds to Einstein's electric photon [8], and a^j is the magnetic four-potential which corresponds to Salam's magnetic photon [9].

Comparison of Eqs. (11) and (12) with Eqs. (4) and (6) demonstrates the analogy between the electric-magnetic duality and the mass-spin duality.

The electric-magnetic duality is required to explain the quantization of electric charge [10]. Quantum field theoretical models which include the magnetic photon can be found in Ref. [11]. I argued [12] that magnetic photon radiation may have already been observed by August Kundt in 1885 [13].

It is probably interesting to note that a Maxwell field A^i which is coupled to a gravitational field which includes both Cartan's torsion [2] and Weyl's non-metricity [14], requires the appearance of the second four-potential a^i [15].

Furthermore, Cartan's torsion tensor can be built from two independent vector fields which appear to obey the modified Maxwell equations of the two-photon theory [16].

3.3 Is There Observational Evidence for Torsion?

The rotation axes of the galaxies of the Perseus-Pisces supercluster are aligned. This alignment exists over a distance of at least 40 Mpc (130 million light years) [16]. Such a large alignment cannot be explained within the framework of conventional models of galaxy-formation. Therefore I suggested [17] that this alignment is either a topological defect (torsion wall [18]) or a remnant of the original aligned distribution of galactic rotation axes generated by a rotating universe [19]. My interpretation of this structure as a torsion wall [20] or as an effect of a rotating universe [21] is now generally accepted.

Bibliography

[1] A. Einstein, *Ann. Phys. (Leipzig)* **49**, 769 (1916).

[2] E. Cartan, *Compt. Rend. Acad. Sci.* **174**, 593 (1922).

[3] T. W. B. Kibble, *J. Math. Phys.* **2**, 212 (1961).
D. W. Sciama, *Rev. Mod. Phys.* **36**, 463 (1964).

[4] F. W. Hehl, P. von der Heyde, G. D. Kerlick, and J. M. Nester, *Rev. Mod. Phys.* **48**, 393 (1976).
F. W. Hehl, J. D. McCrea, E. W. Mielke, and Y. Ne'eman, *Phys. Rept.* **258**, 1 (1995).
I. L. Shapiro, *Phys. Rept.* **357**, 113 (2001).

[5] P. A. M. Dirac, *Proc. Roy. Soc. A* **117**, 610 (1928).

[6] O. Costa de Beauregard, *C. R. Acad. Sci.* **214**, 904 (1942).
O. Costa de Beauregard, *J. Math. Pures Appl.* **22**, 85 (1943).

[7] R. W. Kühne, *Int. J. Mod. Phys. A* **14**, 2531 (1999).

[8] A. Einstein, *Ann. Phys. (Leipzig)* **17**, 132 (1905).

[9] A. Salam, *Phys. Lett.* **22**, 683 (1966).

[10] P. A. M. Dirac, *Proc. R. Soc. A* **133**, 60 (1931).

[11] D. Singleton, *Int. J. Theor. Phys.* **34**, 37 (1995).
D. Singleton, *Am. J. Phys.* **64**, 452 (1996).
D. Singleton, *Int. J. Theor. Phys.* **35**, 2419 (1996).
R. W. Kühne, *Mod. Phys. Lett. A* **12**, 3153 (1997).
S. Carneiro, *J. High Energy Phys.* **9807**, 022 (1998).

[12] R. W. Kühne, *Electromagnetic Phenomena*, **3** (9), 86 (2003) (= hep-ph/0205229).
R. W. Kühne, Has the Last Word Been Said on Classical Electrodynamics?, eds. A. Chubykalo, V. Onoochin, R. Smirnov-Rueda, and A. Espinoza (Rinton Press, New York, to be published).

[13] A. Kundt, *S.-B. Preuß. Acad. Wiss.* (1885) p. 1055.
A. Kundt, *Wied. Ann. Phys. Chem.* **27**, 191 (1886).

[14] H. Weyl, *Ann. Physik* **59**, 101 (1919).

[15] M. Israelit, *Gen. Rel. Grav.* **29**, 1411 (1997).
M. Israelit, *Gen. Rel. Grav.* **29**, 1597 (1997).
M. Israelit, *Found. Phys.* **28**, 205 (1998).
M. Israelit, *Hadronic J.* **21**, 75 (1998).
M. Israelit, The Weyl-Dirac Theory and Our Universe (Nova Science Publishers, New York, 1999).

[16] R. T. Hammond, *Gen. Rel. Grav.* **23**, 973 (1991).

[17] S. A. Gregory, L. A. Thompson, and W. G. Tifft, *Astrophys. J.* **243**, 411 (1981).

[18] R. W. Kühne, *Mod. Phys. Lett. A* **12**, 2473 (1997).

[19] L. C. Garcia de Andrade, *J. Math. Phys.* **39**, 372 (1998).
L. C. Garcia de Andrade, *Mod. Phys. Lett. A* **13**, 1179 (1998).
L. C. Garcia de Andrade, *Phys. Lett. A* **256**, 291 (1999).

[20] K. Gödel, *Rev. Mod. Phys.* **21**, 447 (1949).
A. Raychaudhuri, *Phys. Rev.* **98**, 1123 (1955).

[21] S. Capozziello, G. Iovane, G. Lambiase, and C. Stornaiolo, *Europhys. Lett.* **46**, 710 (1999).
S. Capozziello, G. Lambiase, and C. Stornaiolo, *Ann. Phys. (Leipzig)* **10**, 713 (2001).

[22] A. I. Arbab, *Spacetime & Substance* **2** (7), 55 (2001).
S. Carneiro and G. A. Mena Marugán, *Phys. Rev. D* **64**, 083502 (2001).
C. M. Chen, T. Harko, and M. K. Mak, *Phys. Rev. D* **63** , 104013 (2001).
C. M. Chen, T. Harko, W. F. Kao, and M. K. Mak, *Nucl. Phys. B* **636**, 159 (2002).

In: Relativity, Gravitation, Cosmology
Editor: V. Dvoeglazov and A. Espinoza Garrido, pp. 43-59
ISBN 1-59033-981-9

Chapter 4

On Torsion in a Theory of Gravity

Janusz Garecki
Institute of Physics, University of Szczecin
Wielkopolska 15, 70–451 Szczecin, POLAND
e-mail:garecki@wmf.univ.szczecin.pl

Abstract

It is known that General Relativity (**GR**) uses a Lorentzian Manifold $(M_4; g)$ as a geometrical model of the physical spacetime, that means a four–dimensional differentiable manifold endowed with a Lorentzian metric g. The metric g is required to satisfy Einstein's equations. Since the 1960s many authors have tried to generalize this geometrical model of the physical space–time by introducing torsion into geometrical model of the physical spacetime. In this paper we discuss the present status of torsion in a theory of gravity. At first, we emphasize that up to now we there is no experimental evidence for the existence of torsion in Nature. To the contrary, all experiments performed for a weak gravitational field (Solar System) or for a strong regime (binary pulsars) as well as tests of the Einstein Equivalence Principle (**EEP**) confirmed **GR** and Lorentzian manifold $(M_4; g)$ as satisfactory geometrical model of the physical spacetime. Then, we present some arguments which in our opinion indicate against torsion in gravitational theories. At last, we conclude that the general–relativistic model of the physical spacetime is sufficient for physical applications and it seems to be the most satisfactory.

4.1 Introduction

In the 1960s–1970s some researchers introduced torsion into the theory of gravity [1, 2, 3, 4, 5, 6].[1] The main motives (only theoretical) were the following:

[1] We omit here older attempts to introduce torsion because they have only historical meaning.

1. Analogies between **GR** and the continuum theory of dislocations (Theory of a generalized Cosserat continuum) led to heuristic arguments for a metric spacetime with torsion, i.e., to Riemann–Cartan spacetime.

2. Study of spinning matter in **GR** led some people to the conclusion that the canonical energy–momentum tensor of matter ${}_cT_i{}^k$ is the source of curvature and the canonical intrinsic spin density tensor ${}_cS^{ikl} = (-){}_cS^{kil}$ is the source of torsion of the underlying space–time. From this, Einstein–Cartan–Sciama–Kibble (**ECSK**) theory and its generalizations originated.

3. Attempts to formulate theory of gravity as a gauge theory for Lorentz group L or for Poincare' group P led to a space–time endowed with a metric– compatible connection which might have non–vanishing torsion, i.e., again one was led to Riemann–Cartan space–time [7, 8, 9, 10, 11].

The above theoretical motives *are not convincing.*[2] For example, the often used argument for torsion (following from 2.) based on the (non-homogeneous) holonomy theorem [4, 11][3] holds only if one uses *Cartan displacement* which displaces vectors and contactpoints [12]. Ordinary parallel displacement (which displaces only vectors) gives only Lorentz rotations (= homogeneous holonomy group) even in a Riemann-Cartan space-time [12]. Moreover, there are other geometrical interpretations of torsion, e.g., Bompiani [13] *connects torsion with rotations in tangent spaces*, not with translations.

We also needn't to generalize **GR** in order to get a gauge theory with L or P as a gauge group [14, 47]. Note, in particular, that *Asthekar's approach* to **GR** [15, 16, 17, 18] leads to a gauge theory of gravity akin to a Yang–Mills theory.

Up to now, *there is no experimental evidence for the existence of torsion in Nature.* To the contrary, all experiments confirmed with a very high precision ($\sim 10^{-17}$) Einstein's Equivalence Principle (**EEP**) and, with a smaller precision ($\sim 99.9\%$) the General Relativity (**GR**) equations [19, 20, 21, 22]. Here by **EEP** we mean a formulation of this Principle given by C. W. Will [19, 20]. In this formulation[4] the **EEP** states:

1. The Weak Equivalence Principle (**WEP**) is valid. This means that the trajectory of a freely falling spherical test body (one not acted upon by non-gravitational forces and being too small to be affected by tidal forces) *is independent of its internal structure and material composition.*

---

[2] The Referee's suggested the two new motives: 1. the duality between curvature and torsion, 2. thermodynamically, the concept of torsion is necessary ingredient of non-equilibrium non-linear processes. I must say that I cannot understand what exactly means this "duality between curvature and torsion"? I suppose that Referee have thought about "teleparallel equivalence of **GR**". I know that there are trials to formulate the classical thermodynamics in terms of Riemannian or more general Finsler geometry. But it is other problem, not connected with a theory of gravity, i.e., with the spacetime structure. I also do not think that torsion is needed to explain Universe expansion, Sagnac effect or, why a plane is flying (as one of the Referee claimed).

[3] This theorem says that torsion gives translations, and curvature gives Lorentz rotations in tangent spaces of a Riemann-Cartan manifold induced by (Cartan) displacement along loops.

[4] In this formulation the Principle can be experimentally tested.

2. Local Lorentz Invariance (**LLI**) is valid. This means that the outcome of any local non-gravitational experiment *is independent of the velocity of the freely-falling and non-rotating reference frame in which it is performed.*

3. Local Position Invariance (**LPI**) is valid. This means that the outcome of any local non-gravitational experiment performed in a freely-falling and non-rotating reference frame *is independent of where and when in the Universe it is performed.*

The only theories of gravity that can embody **EEP** in the above constructive formulation of the Principle are those that satisfy the postulates of *metric theories of gravity* [19, 20], which are:

1. Spacetime is endowed with a symmetric metric.

2. The trajectories of freely falling spherical test bodies are the geodesics of that metric.

3. In local freely-falling and non-rotating reference frames the non-gravitational laws of physics are those written in the language of Special Relativity (**SRT**).

The **EEP** implies a *universal pure metric coupling between matter and gravity.* It admits **GR**, of course, and, at most, some of the so–called *scalar–tensor theories* (these, which respect **EEP**) [20, 21, 22] without torsion.

So, torsion seems to be *excluded* in vacuum or at least *very strongly constrained* in vacuum by the latest gravitational experiments , i.e., at least propagating torsion is excluded or very strongly constrained by these experiments)[5] which have confirmed **EEP** with very high precision. As a consequence, at least propagating torsion still seems to be *purely hypothetical.*

The paper is organized as follows. In Section II we pay our attention to some characteristic properties of torsion and spacetimes with torsion. We will conclude from these properties that torsion rather *should not be introduced into a theory of gravity.*

The consideration about energy–momentum tensor for matter given in Section III also support the view that the torsion *is not needed* in a theory of gravity.

In Section IV we give some remarks about the so–called "teleparallel equivalent of general relativity" (**TEGR**) and in Section V we give some conclusions.

In the paper we will confine to the metric–compatible connection $\omega^i{}_k$. We cannot see any reasons to consider more general connections, for example, the connection proposed in [23, 24, 25].

[5] Torsion is excluded or very strongly constrained at least in vacuum because if we neglect a cosmological background, then the all gravitational experiments were performed in vacuum. This means that **ECSK** theory can survive since this theory is identical in vacuum with **GR**. Of course, the same is true for other gravity theories which in vacuum reduce to **GR**.

For a metric–compatible connection ${\omega^i}_k$ we have

$$Dg = 0 \tag{4.1}$$

and

$${\omega^i}_k =_{LC} {\omega^i}_k + {K^i}_k, \tag{4.2}$$

where ${}_{LC}{\omega^i}_k$ denotes the *Levi–Civita connection* and ${K^i}_k$ denotes *contortion.*

By $\vartheta^i := \vartheta^0, \vartheta^1, \vartheta^2, \vartheta^3$ we will denote a Lorentzian coreper which is defined by the following property: the metric g in this coreper has the form $g = \eta_{ik}\vartheta^i \bigotimes \vartheta^k$, where η_{ik} is the Minkowskian metric. Ψ will denote a matter field and D will mean the exterior covariant derivative. For tensor fields the exterior covariant derivative D reduces to an ordinary covariant differential $D = \vartheta^i \nabla_i$, where ∇_i means covariant derivative.

4.2 Some characteristic properties of torsion and spacetimes with torsion

We begin this Section with the remark that if one utilizes the so–called "Ockham's razor"[6] then torsion is needn't for him in a theory of gravity because the *wonderful, the most simple and most symmetric* Levi–Civita connection is sufficient for the all physical requirements.

Now, let us pay our attention to the following, special properties of torsion and metric spacetimes with torsion:

1. In a spacetime with torsion *do not exist* infinitesimal parallelograms [12, 26] because the operation of invariant geometric addition of infinitesimal coordinate segments is noncommutative. So, such spacetimes seem *physically inadmissible* as this result is in direct conflict with the operational and epistemological basis of our difference physics [27].

2. *Torsion is topologically trivial.* This means that the topological invariants of a real manifold M and characteristic classes of vector bundles over M, as defined in [28, 29, 30] depend only on curvature and can be fully determined by the curvature ${}_{LC}{\Omega^i}_k$ of the Levi–Civita connection. Roughly speaking, one can continuously deform any metric–compatible connection (or even general linear connection) into Levi-Civita connection *without changing topological invariants and characteristic classes.* So, torsion is not relevant for topological invariants and characteristic classes.[7]

[6] By "Ockham's razor" we mean a Philosophical Principle which states: "Entities are not to be multiplied without necessity".

[7] Some authors say that torsion which satisfies some differential field equations might be topologically non-trivial. But this seems to be incorrect because one can still continuously deform the connection in the case into torsionless Levi-Civita connection *without changing topological invariants and characteristic classes.* The field equations may, of course, change during such deformation. So, it seems to us that one can say only that the torsion which

3. *Torsion is not relevant from the dynamical point of view either.* Namely, one can reformulate every metric theory of gravitation with a metric- compatible connection ${\omega^i}_k$ as a "Levi-Civita theory". Torsion is then treated as *a matter field.* Such reformulation preserves the all dynamical properties of the theory. An obvious example is given by **ECSK** theory in the so–called "combined formulation" [31].[8] In general, one can prove [33] that any total Lagrangian of the type

$$L_t = L_g(\vartheta^i, {\omega^i}_k) + L_m(\Psi, D\Psi) \tag{4.3}$$

admits an unique decomposition into a pure geometric part $\tilde{L}_g(\vartheta^i, {}_{LC}{\omega^i}_k)$ containing no torsion plus a generalized matter Lagrangian $\tilde{L}_m(\Psi, {}_{LC}D\Psi, {K^i}_k, {}_{LC}D{K^i}_k)$ which collects the pure matter terms and all the terms involving torsion

$$L_t = L_g + L_m = \tilde{L}_g + \tilde{L}_m. \tag{4.4}$$

Here ${}_{LC}D$ means the exterior covariant derivative with respect to the Levi-Civita connection ${}_{LC}{\omega^i}_k$.

From the Lagrangian

$$L_t = \tilde{L}_g + \tilde{L}_m \tag{4.5}$$

there follow the *Levi-Civita equations associated with* L_t.

So, torsion *can always be treated as a matter field.* This point of view is preferred e.g. in [34, 35, 36, 37] and it is supported by transformational properties of torsion: torsion transforms like a matter field i.e., it transforms as a tensor–valued form.

4. In a spacetime with torsion a tangent space $T(P)$ cannot be identified with Minkowskian spacetime because there do not exist coordinates [9] such that $g_{ik}(P) = \eta_{ik}$, $\Gamma^i_{kl}(P) = 0$. So, a gravitational theory with torsion *is not a covering theory for* **SRT** [57] and, therefore, it violates **EEP**. We also lose Fermi coordinates [12, 38, 39] in a Riemann-Cartan space-time.[10]

Some authors [40, 41, 42, 43] assert that the **EEP** can be realized in some sense also in Riemann–Cartan geometry because there can be introduced an *anholonomic tetrad* at any point P in which ${\Gamma^i}_{kl}(P) = 0$, $g_{ik}(P) = \eta_{ik}$. But the proofs of this *anholonomic formulation* of the **EEP** given in [40, 41, 42] and repeated in [43] are incomplete because there was not given

---

satisfies some differential field equations *might be physically* non-trivial. Of course, we cannot exclude that there exist other topological properties of spacetime which can substantially depend on torsion, e.g., following one of the Referee torsion is non zero in domains of topological dimension 3 or more. But we think that he had in mind the *topological torsion* instead of *differential-geometric* torsion.

[8] In this formulation **ECSK** theory is dynamically fully equivalent to the ordinary **GR** [32].

[9] By "coordinates" we mean what differential geometrists mean, i.e., *local maps.* They are always holonomic.

[10] Fermi coordinates realize in **GR** a local (freely-falling and non-rotating) inertial frame along a curve in which **SRT** is valid.

any explicite solution of the strongly overdetermined systems of partial differential equations used in there. The correct, very general proof, which however does not use transformational rule for connection, is given in [44]. But the proof given in [44] is valid for *any linear connection* in a metric spacetime; not only for the metric compatible connection. So this proof shows that the above mentioned *anholonomic formulation* of the **EEP** *has no operational meaning because it is trivially satisfied* in any metric manifold endowed with a linear connection.

Constructive formulation of the **EEP** needs holonomic coordinates (See e.g. [19, 38, 45, 62]). For example the Equivalence Principle formulated in a holonomic coordinates in the same form as it was formulated above in an anholonomic frame *effectively works* and admits only a symmetric linear connection. Then, adding the most natural *metricity postulate* (or Hamiltonian Principle for trajectories of the test particles) unequivocally leads us to (pseudo)-Riemannian geometry, i.e., to the Levi-Civita connection.

5. A connection having torsion can be determined neither by its own autoparallels (paths) nor by geodesics [12]. So, one cannot determine unequivocally a connection which has torsion by observation of the test particles (which could move along geodesics or autoparallels).

6. The gravitational theories with metric–compatible torsion, apart from **ECSK** theory, have *much weaker* field equations in Einsteinian sense than do the competitive purely metric gravity theories [46]. Namely, the gravitational theories with metric–compatible torsion have much greater *Einsteinian strength* of their field equations than the purely metric gravity theories, i.e., these theories *determine gravitational field with much smaller precise than the purely metric gravity theories do* [46].

Torsion leads to ambiguities:

1. The *Minimal Coupling Principle* (**MCP**) differs from the *Minimal Action Principle* (**MAP**) in a spacetime with torsion [47].

   The **MCP** can be formulated as follows. In **SRT** field equations obtained from the **SRT** Lagrangian density $L = L(\Psi, \partial_i\Psi)$ we replace $\partial_i \longrightarrow \nabla_i, \ \eta_{ik} \longrightarrow g_{ik}$ and get *covariant field equations* in (M_4, g).

   By the **MAP** we mean an application of the Minimal Action Principle (Hamiltonian Principle) to the covariant action integral $S = \int\limits_{\Omega} L(\Psi,\ D\Psi)\, d^4\Omega$, where $L(\Psi,\ D\Psi)$ is a covariant Lagrangian density obtained from the **SRT** Lagrangian density $L(\Psi,\ \partial_i\Psi)$ by **MCP**.

   It is natural to expect that the field equations in (M_4, g) obtained by using **MCP** on **SRT** equations should coincide with the Euler-Lagrange equations obtained from $L(\Psi,\ D\Psi)$ by **MAP**. *This holds in GR but not* *in the framework of the Riemann-Cartan geometry.* So, we have there an ambiguity of the field equations.[11]

2. In the framework of the **ECSK** theory of gravity we have four energy-momentum tensors for matter: Hilbert, canonical, combined, formal [31]. Which one is correct?

3. Let us consider now normal coordinates **NC(P)** [12, 48, 49, 50] which are so very important in **GR** (See, eg., [48, 49, 50, 51]). In the framework of the Riemann-Cartan geometry we have two **NC(P)**: normal coordinates for the Levi-Civita part of the Riemann-Cartan connection $\mathbf{NC}({}_{\mathbf{LC}}\omega,\ \mathbf{P})$ and normal coordinates for the symmetric part of the full connection $\mathbf{NC}({}_{\mathbf{s}}\omega,\ \mathbf{P})$ [52, 53]. Which one has a greater physical meaning?

   The above ambiguity of the normal coordinates[12] leads us to ambiguities in superenergy and supermomentum tensors [52, 53, 54]. Moreover, the obtained expressions are too complicated for practical use. In fact, we lose here a possibility of effective use of the normal coordinates.

   Perhaps by use *normal frames* defined in [44] instead of normal coordinates one could avoid these ambiguities. This problem will be studied in future.

4. In the framework of Riemann-Cartan geometry [12] there holds

$$R_{(ik)lm} = R_{ik(lm)} = 0, \tag{4.6}$$

   but

$$R_{iklm} \neq R_{lmik}. \tag{4.7}$$

   The last asymmetry leads to an ambiguity in construction of the so–called "Maxwellian superenergy tensor" for the field R_{iklm} [54, 55]. This tensor is uniquely constructed in **GR** owing to the symmetry $R_{iklm} = R_{lmik}$ and it is proportional to the Bel-Robinson tensor [52, 53, 54, 55, 56]. In the framework of the Riemann-Cartan geometry the obtained result depends on which antisymmetric pair of the R_{iklm}, the first or second, is used in the construction.

5. In a Riemann-Cartan spacetime we have geodesics and autoparallels (paths). Hamiltonian Principle demands geodesics as trajectories for the test particles [57]. Then, what about the physical meaning of the autoparallels?[13]

6. In a spacetime with torsion we have in fact two kinds of parallel displacement defined by

$$dv^k = (-)\Gamma^k_{ij} v^j dx^i \tag{4.8}$$

[11] Axial torsion removes this ambiguity. By (M_4, g) we mean here a general metric manifold; not necessarily Riemannian.

[12] Axial torsion removes this ambiguity.

[13] Axial torsion removes this problem. One can also easily prove in the framework of the **ECSK** theory that spinless test particles move along geodesics.

and

$$dv^k = (-)\Gamma^k_{ij} v^i dx^j. \tag{4.9}$$

These results follow from that two kinds of absolute (and covariant) differentials

$$\nabla_i v^k = \partial_i v^k + \Gamma^k_{il} v^l \tag{4.10}$$

$$\nabla_i v^k = \partial_i v^k + \Gamma^k_{li} v^l. \tag{4.11}$$

Which one of the two above possibilities could be eventually realized in Nature?

In practice, one must consequently use one of the two above possibilities (or conventions) during any calculations in order to avoid mistakes. For example, in Hehl's papers

$$\nabla_i \sqrt{|g|} = 0 \tag{4.12}$$

in the Riemann-Cartan space-time (which is, of course, correct result). But in the paper [58] one can find

$$\nabla_i \sqrt{|g|} \neq 0. \tag{4.13}$$

The last result is, of course, not correct and it is a consequence of mixing of the two formulas for covariant differentiation.

Another source of computational mistakes connected with torsion is the following: different authors use definitions of torsion which differ by sign or by a factor 1/2.

4.3 Symmetry of the energy–momentum tensor of matter

In Special Relativity (**SRT**) the correct energy–momentum tensor for matter (electromagnetic field,continuous medium, dust, elastic body, solids) *must be symmetric* [38, 59]. One can always get such a tensor starting from the *canonical pair* ${}_cT^{ik}, {}_cS^{ikl} = (-)_cS^{kil}$, where ${}_cT^{ik} \neq_c T^{ki}$ is the canonical energy-momentum tensor and ${}_cS^{ikl}$ — the canonical spintensor. These two canonical tensors are connected by the equations

$$\partial_k {}_cT^{ik} = 0, \quad {}_cT^{ik} -_c T^{ki} = \partial_l S^{ikl}. \tag{4.14}$$

By use of the *Belinfante symmetrization procedure* [31, 47, 60, 61] one can get the most simple new pair

$${}_sT^{ik} =_c T^{ik} - \frac{1}{2}\partial_j\left({}_cS^{ikj} -_c S^{ijk} +_c S^{jki}\right), \tag{4.15}$$

$$S^{ijk} =_c S^{ijk} - A^{jki} + A^{ikj} = 0. \tag{4.16}$$

Here

$$A^{ikj} = \frac{1}{2}\left({}_cS^{ikj} -_c S^{ijk} +_c S^{jki}\right). \tag{4.17}$$

The obtained new "pair" (${}_sT^{ik}$, 0) is *the most simple and the most symmetric.* Note that the symmetric tensor ${}_sT^{ik} =_s T^{ki}$ gives complete description of matter because the spin density tensor ${}_cS^{ijk}$ is *entirely absorbed* into ${}_sT^{ik}$ by the symmetrization procedure.

It is interesting that one can easily generalize the above symmetrization procedure onto a general metric manifold (M_4, g) [14, 31] by using the Levi-Civita connection associated with the metric g. The generalized symmetrization procedure has the same form as above with the replacement $\eta_{ik} \longrightarrow g_{ik}, \ \partial_i \longrightarrow_{LC} \nabla_i$.

So, one can always get on a metric manifold (M_4, g) a symmetric energy–momentum tensor ${}_sT^{ik} =_s T^{ki}$ for matter (then, of course, corresponding $S^{ikj} = 0$). Observe that the symmetric tensor ${}_sT^{ik}$, like as in **SRT**, consists of the canonical tensors ${}_cT^{ik}$ and ${}_cS^{ikl}$.

The symmetric energy–momentum tensor for matter is uniquely determined by the matter equations of motion and reasonable boundary conditions [62]. This fact is very important for the uniqueness of the gravitational field equations. Moreover, the symmetric energy–momentum tensor is covariantly conserved (a canonical energy-momentum tensor is not conserved).

L. Rosenfeld has proved [63] that

$${}_sT^{ik} = \frac{\delta L_m}{\delta g_{ik}}, \tag{4.18}$$

where $L_m = L_m(\Psi, \ {}_{LC}D\Psi)$ is a covariant Lagrangian density for matter. The tensor ${}_sT^{ik}$ given by (18) is the source in the Einstein equations

$$G_{ik} = \chi_s T_{ik}, \tag{4.19}$$

where $\chi = \frac{8\pi G}{c^4}$.

Note that these equations *geometrize both the canonical quantities* ${}_cT^{ik}$ and ${}_cS^{ikl} = (-)_cS^{kil}$ *in some equivalent way* because the tensor ${}_sT^{ik}$ is built from these two canonical tensors.

So, it is the most natural and most simple to postulate that, in general, the correct energy–momentum tensor for matter is the symmetric tensor ${}_sT^{ik}$. This leads us to a purely metric torsion-free theory of gravity with the field equations

$$\frac{\delta L_g}{\delta g_{ik}} = \frac{\delta L_m}{\delta g_{ik}}. \tag{4.20}$$

Then, if we take into account the *dynamical universality* of the Einstein equations [35, 36, 37, 64, 65, 66, 67], we will end up with General Relativity (possibly with $\Lambda \neq 0$) which will have a sophisticated, symmetric energy-momentum tensor as a source.

4.4 Remarks on the "teleparallel equivalent of general relativity"

After presenting the preliminary draft of the paper as e–print, we have got critical remarks from some persons which are working on the so-called *teleparallel equivalent of general relativity* (**TEGR**) in the framework of the *Weitzenböck or teleparallel geometry* [12, 26, 68, 69, 70].

The *Weitzenböck or teleparallel connection and geometric structure* on spacetime is determined by a tetrad (or other anholonomic frame) field $h^{(a)}{}_b(x)$ and *can always* be introduced *independently* of the primordial geometric structure of the spacetime. Here $(a), (b), \ldots$ are tetrad (= anholonomic) indices and $a, b, c, \ldots$ mean holonomic (= world) indices.

The fundamental formulas of the teleparallel geometry read

$$g_{ik} := \eta_{(a)(b)} h^{(a)}{}_i h^{(b)}{}_k, \tag{4.21}$$

$$\Gamma^i{}_{kl} := h_{(a)}{}^i \partial_k h^{(a)}{}_l, \tag{4.22}$$

$$\nabla_i h^{(a)}{}_k = 0, \tag{4.23}$$

$$\Gamma^i{}_{kl} =_{LC} \Gamma^i{}_{kl} + K^i{}_{kl}, \tag{4.24}$$

$$K^i{}_{kl} := 1/2\big(T_k{}^i{}_l + T_l{}^i{}_k - T^i{}_{kl}\big), \tag{4.25}$$

$$T^i{}_{kl} := \Gamma^i{}_{lk} - \Gamma^i{}_{kl}, \tag{4.26}$$

and

$$R^i{}_{klm} =_{LC} R^i{}_{klm} + Q^i{}_{klm} \equiv 0, \tag{4.27}$$

where $Q^i{}_{klm}$ is a tensor written in terms of the *contortion* $K^i{}_{kl}$ and its covariant derivatives with respect to the Levi-Civita connection ${}_{LC}\Gamma^i{}_{kl}$ of the metric g_{ik}.

Here $\eta_{(a)(b)}$ means the *interior metric* (usually Minkowskian) of a tangent space and the duals $h_{(a)}{}^i$ are defined by

$$h_{(a)}{}^i h^{(a)}{}_k = \delta^i_k. \tag{4.28}$$

Those authors which work on **TEGR**, by use the formulas (24) and (27) of the teleparallel geometry *rephrase*, step-by-step, the all formalism of **GR** in terms of the Weitzenböck connection $\Gamma^i{}_{kl}$ and its *torsion* $T^i{}_{kl}$. Then, they call this formal reformulation of **GR** in terms of the Weitzenböck geometry *the teleparallel equivalent of general relativity* (**TEGR**) (What kind of "equivalence"?).

One can read in the papers [68, 69, 70] the following conclusion: "Gravitational interaction, thus, can be described alternatively in terms of curvature, as is usually done in **GR**, or in terms of torsion, in which case we have the so–called *teleparallel gravity*. Whether gravitation requires a curved or torsional spacetime, therefore, turns out to be a matter of convention".

From the point of view of the **TEGR**, therefore, teleparallel torsion has fundamental physical meaning and it has been already detected.

We cannot agree with such statements. In our opinion, the "teleparallel equivalent of **GR**" is only *formal* and geometrically *trivial* rephrase of **GR** in terms of the Weitzenböck geometry. Such rephrase is, of course, *always possible* not only with **GR** but also with any other purely metric theory of gravity (see eg. [71]) but it *has no profound physical motivation.* It is because, as one can easily show, the teleparallel torsion can be entirely expressed in terms of the Van Danzig and Schouten aholonomity object $\Omega^{(a)}{}_{(b)(c)}$ (see eg. [12, 26]). So, the torsion $T^i{}_{kl}$ of a teleparallel connection *describes only anholonomity* of the used field of aholonomic frames $h^{(a)}{}_i(x)$; *not real geometry of the spacetime.*[14] Only Levi-Civita part of a Weitzenböck connection can have (and has) physical meaning.

Resuming, it seems to us that **TEGR** is rather a *mathematical curiosity* which gives, by no means, anything better than ordinary **GR** and one can doubt into its physical meaning.

Precise experimental confirmation of the **EEP** proved non-zero curvature of physical spacetime [38, 77] and supported ordinary **GR**. We think that this fact excludes a physical motivation for *rephrasing* **GR** into **TEGR**.

4.5 Concluding remarks

The **GR** model of the space-time has very good experimental confirmation in a weak-field approximation (Solar System) and in the strong fields (binary pulsars). On the other hand, torsion has no experimental evidence (at least in vacuum) and it is not needed in a theory of gravity. Moreover, the introduction of torsion into the geometric structure of space-time leads to many problems (apart from calculational, of course). Most of these problems are removed if only axial torsion $A_i = \frac{1}{6}\eta_{iabc}Q^{abc}$, $Q^{[abc]} = Q^{abc}$ exists. So, it would be reasonable to confine themselves to the axial torsion only (If one still want to keep on torsion). This also could be supported by the important fact that the *matter fields* (= Dirac's particles) are coupled to the axial part only of torsion in the Riemann-Cartan space-time. However, if we confine to the axial torsion, then (if we remember the dynamical triviality of torsion and the dynamical universality of the Einstein equations) we effectively will end up with **GR** + additional matter fields. In the most important case of the **ECSK** theory we will end up with **GR** + an additional pseudovector field A_i (or with an additional pseudoscalar field φ if the field A_i is potential, i.e., if $A_i = \partial_i\varphi$) [47]. But **GR** with an additional dynamical pseudovector field A_i yields local gravitational physics which may have both location and velocity-dependent effects [20] unobserved up to

---

[14] Unless one can physically distinguish a tetrad field (or other anholonomic field of frames) and give it a fundamental geometrical and physical meaning. But we think that this could introduce a crystal–like structure on spacetime and, therefore, it would contradict local Lorentz invariance.

now, and **GR** with an additional pseudoscalar field has a defect because there exist two distinguished frames, *the Einstein frame* and *the Jordan frame*, which are not equivalent physically [78].

So, we will finish with the conclusion that the geometric model of the spacetime given by ordinary **GR** and "wonderful" Levi-Civita connection *seems to be the most satisfactory*[15].

It seems to us that the torsion was introduced into a theory of gravity in order to get some link between theory of gravity and quantum fundamental particles theory (It is commonly known that the role of the curvature in an atomic and smaller scale is negligible). But these trials *were not successful.* It also seems that what we really need nowadays is a *quantum model* of the Riemannian geometry and a *quantum gravity* which is based on this model. The recent papers given by Ashtekhar [16, 17, 18] and co–workers on this problem seems to be very promising.

Acknowledgment

The author would like to thank Prof. F.W. Hehl, Prof. A. Trautman and Prof. M. Blagojevič for their, most useful, critical remarks. Especially, he would like to thank Prof. Rainer Schimming for his interest to this paper and for his many useful discussions and hints. Owing to these discussions and hints the paper was essentially amended.

[15]During the Symposium LISA 3 (Golm, July 10-14, 2000) I have discussed with many Participiants of the Symposium about the proper geometric model of spacetime. They all agreed with the above conclusion. See also the conclusion about the correct theory of gravity and spacetime structure presented in [79].

Bibliography

[1] R. Utiyama, *Phys. Rev.*, **101** (1956) 1597.

[2] T.W.B. Kibble, *Journ. Math. Phys.*, **2** (1961) 212.

[3] D.W. Sciama,"On the analogy between charge and spin in general relativity" in *Recent developments in general relativity*, **PWN, Warsaw 1962**.

[4] A.Trautman, "On the Structure of the Einstein–Cartan Equations", *Istituto Nazionale di Alta Matematica, Symposia Mathematica*, **12** (1973) 139.

[5] F.W.Hehl, *Gen. Relativ. Grav.*, **4** (1973) 333; –ibidem **5** (1974) 491.

[6] F.W. Hehl et al., *Rev. Mod. Phys.* **48** (1976) 393.

[7] F. W. Hehl et al., "Gravitation and the Poincare' Gauge Field Theory with Quadratic Lagrangian" in *General Relativity and Gravitation* Vol. I. Ed. A. Held, **Plenum Publishing Corporation 1980**.

[8] A. Trautman, "Fiber Bundles, Gauge Fields and Gravitation" in *General Relativity and Gravitation* Vol. I. Ed. A. Held, **Plenum Publishing Corporation 1980**.

[9] K. Hayashi and T. Shirafuji, *Prog. Theor. Phys.*, **64** (1980) 866; **64** (1980) 883; **64** (1980) 1435; **64** (1980) 2222; **65** (1981) 525; **66** (1981) 318; **66** (1981) 2258; **73** (1985) 54; **74** (1985) 852.

[10] J. Garecki, *Gen. Relativ. Grav.*, **22** (1990) 111.

[11] A. Trautman, *Class. Quantum Grav.* **16** (1999) A157-A176.

[12] J.A. Schouten, "Ricci-Calculus", **Springer-Verlag, Berlin 1954**.

[13] E. Bompiani, "Significato del tensore di torsione in une connesione affine", *Boll, Un. Mat. Ital.* **6** (1951) 273.

[14] D. Grensing and G. Grensing, *Phys. Rev.,* **D 28** (1983) 286.

[15] D. Giulini, "Ashtekar Variables in Classical General Relativity" in *Canonical Gravity: From Classical to Quantum*, Eds. J. Ehlers and H. Friedrich, **Springer–Verlag 1994**.

[16] A. Ashtekar, "Quantum Mechanics of Geometry", e–print gr-qc/9901023.

[17] A. Ashtekhar, *Ann. Phys.(Leipzig)*, **9** (2000) 178. E-print gr-qc/99101101 27 Oct 1999.

[18] M.S. Iriondo et al., *Adv. Theor. Math. Phys.*, **2** (1988) 1075. E-print gr-qc/9804019 v4 31 Mar 1999.

[19] Clifford M. Will, *Theory and experiment in gravitational physics*, **Cambridge University Press, Cambridge 1993**.

[20] Clifford M. Will, "The Confrontation Between General Relativity and Experiment: e–print gr-qc/0103036.

[21] G. Esposito-Farese, "Comparing Solar-System, Binary-Pulsars, and Gravitational-Wave Test of Gravity", *e–print gr-qc/9903058.*

[22] T. Damour, "Experimental Tests of Relativistic Gravity", *e–print gr-qc/9904057.*

[23] F.W. Hehl, J.D. McCrea, E.W. Mielke and Y. Neéman, *Found. Phys.*,**19** (1989) 1075.

[24] F.W. Hehl et al., *Phys. Rep.* **258** (1995) 1–171.

[25] F. Gronwald, "Metric–Affine Gauge Theory of Gravity I. Fundamental Structure and Field Equations", *e–print gr-qc/9702034* 18 Feb 1998.

[26] S. Gołąb, "Tensor Calculus", **PWN** (Warsaw 1966) (in Polish). [Exists English translation].

[27] D.G.B. Edelen, *The structure of field space*, **Berkeley , University of California Press 1962**.

[28] A. Trautman, "Differential Geometry for Physicists", *Stony Brook Lectures*, **Bibliopolis, Naples 1984**.

[29] S. Kobayashi and K. Nomizu, *Foundations of Differential Geometry*, **Interscience Publishers 1963 (Vol.I) and 1969 (Vol.II)**.

[30] K. Maurin, two articles in *Leksykon Matematyczny.* Eds. M. Skwarczyński, **Wiedza Powszechna, Warsaw 1993** (in Polish).

[31] F. W. Hehl, *Rep. Math. Phys.*, **9** (1976) 55.

[32] J. Nester, *Phys. Rev.* **D 16** (1977) 239.

[33] M. A. Schweitzer, *Gauge Theory and Gravitation*, **PhD Zürich 1980**.

[34] M. Ferraris, J. Kijowski, *Gen. Rel. Grav.*, **14** (1982) 37.

[35] A. Jakubiec, K. Kijowski, *Phys. Rev.*, **D 37** (1988) 1406.

[36] A. Jakubiec, J. Kijowski, *J. Math. Phys.* **30** (1989) 1073.

[37] A. Jakubiec, J. Kijowski, *J. Math. Phys.*, **30** (1989) 2923.

[38] C. W. Misner, K. S. Thorne, J. A. Wheeler, *Gravitation*, **W. H. Freeman and Company, San Francisco 1973**.

[39] Wann-Quan Li and Wei-Tou Ni, *J. Math. Phys.*, **20** (1979) 25.

[40] P. von der Heyde, *Lett. Nuovo Cim.*, **14** (1975) 250.

[41] F.W. Hehl, "Four lectures on Poincare' Theory of Gravity", *Proceedings of the 6th Course of the International School of Cosmology and Gravitation on "Spin, Torsion, Rotation and Supergravity"* held at Erice, Italy, May 1979. P.G. Bergmann and V. de Sabatta (editors), Plenum Press 1980.

[42] F.W. Hehl, *On the Gauge Field Theory of Gravitation*, 3 lectures given at the Dublin Institute for Advanced Studies from 18–21 Nov. 1981.

[43] M. Blagojević, *Gravitation and Gauge Symmetries*, **Institute of Physics Publishing, Bristol 2002**.

[44] D. Hartley, *Class. Quantum Grav.*, **12** (1995) L103-L106. E-print Archive: gr-qc/9510013.

[45] L.D. Landau, E.M. Lifschitz, *Field Theory*, **Nauka Publishers, Moscow 1988** (In Russian) [English edition of this book: "The Classical Theory of Fields" Pergamon, Oxford 1975].

[46] J. Garecki, R. Schimming, *Rep. Math. Phys*, **51** (2003) 197. E-print gr-qc/0201008.

[47] V. de Sabatta and M. Gasperini, *Introduction to Gravitation*, **World Scientific, Singapore 1985**.

[48] A.Z. Petrov, *New Methods in General Relativity*, **Nauka, Moscow 1966** (In Russian). [English edition of Petrov's book: "Einstein Spaces", Pergamon, New York 1969].

[49] L. Brewin, *Class. Quantum Grav.*, **25** (1998) 3085.

[50] A.I. Nesterov, *Class. Quantum Grav.*, **16** (1999) 465.

[51] V. Müller et al., "A Closed Formula for the Riemann Normal Coordinate Expansion", *e–print gr-qc/9712092* v2.

[52] J. Garecki, *Acta Phys. Pol.*, **B 9** (1978) 291.

[53] J. Garecki, *Acta Phys. Pol.*, **B 12** (1981) 739.

[54] J. Garecki, *Class. Quantum Grav.*, **2** (1985) 403.

[55] J. Garecki, *Acta Phys. Pol.*, **B 12** (1981) 1017.

[56] J. Garecki, "Some Remarks on the Bel-Robinson Tensor". An amended version of the Essay which received an "honorable mention" in the 1999 Essay Competition of the Gravity Research Foundation. E–print gr-qc/0003006. *Ann. Phys. (Leipzig)* **10** (2001) 911.

[57] F. Rohrlich, *Classical Charge Particles. Foundations of their Theory,* **Addison-Wesley Publishing Company, Inc. Reading, Massachusets 1965**.

[58] A. Saa, *Gen. Rel. Grav.*, **29** (1997) 205.

[59] B. F. Schutz, *A First Course in General Relativity,* **Cambridge University Press, Cambridge 1985**.

[60] F. J. Belinfante, *Physica,* **6** (1939) 887.

[61] W. Kopczyński, A. Trautman, *Spacetime and Gravitation,* **PWN, Warsaw 1984** (in Polish). [English edition of this book: "Space-Time and Gravitation", J. Wiley, New York 1992].

[62] V. D. Fock, *The Theory of Space, Time and Gravitation,* **Pergamon Press, London 1969**.

[63] L. Rosenfeld, *Mem. Acad. Roy. Belg.*, **Cl. Sc., fasc. 6** (1940).

[64] G. Magnano et al., *Gen. Rel. Grav.*, **19** (1987) 465.

[65] A. Borowiec et al., *Gen. Rel. Grav..*, **26** (1994) 637.

[66] A. Borowiec et al., *Class. Quantum Grav.*, **15** (1998) 43 (gr-qc/9611067).

[67] A. Borowiec et al., "Alternative Lagrangians for Einstein Metrics", *e–print gr-qc/9806116* v2

[68] V.C. de Andrade et al., "Teleparallel gravity: an overview", *e–print gr-qc/0011087.*

[69] V.C. de Andrade et al., "Teleparallel gravity and the gravitational energy-momentum density", *e–print gr-qc/0011097.*

[70] V.C. de Andrade et al., "Teleparallel spin connection", *e–print gr-qc/0104102.*

[71] J. Garecki, R. Schimming, *Bulletin dela Societe' des Sciences et des Lettres de Łódź,* **Vol. LII** (2002) 101.

[72] M. Blagojevič and M. Vasilič, "Conservation laws in the teleparallel theory of gravity", *e–print hep-th/0008159.*

[73] G.Y. Chee et al., "Gravitational energy-momentum and the Hamiltonian formulation of the teleparallel gravity", *e–print gr-qc/0106053.*

[74] M. Blagojevič, "Three Lectures on Poincare' Gauge Theory", *e-print gr-qc/0302040.*

[75] R.T. Hammond, *Torsion Gravity*, draft, September 25, 2001.

[76] Yuyui Lam, "Totally asymmetric torsion on Riemann-Cartan manifold", *e-print gr-qc/0211009.*

[77] A. Schild, "Lectures on General Relativity Theory" in *Relativity Theory and Astrophysics.* Ed. J. Ehlers, **AMS, Providence, Rhode Island, 1967**; "Gravitational Theories of Whitehead Type and the Principle of Equivalence" in *Evidence for Gravitational Theories.* Ed. C. Møller, **Academic Press, New York 1962**.

[78] A. Macias and A. Garcia, *Gen. Relativ. Grav.*,**33** (2001) 890.

[79] N. Straumann, "Reflections on Gravity", *e–print astro-ph/0006423.*

In: Relativity, Gravitation, Cosmology
Editor: V. Dvoeglazov and A. Espinoza Garrido, pp. 61-69
ISBN 1-59033-981-9

Chapter 5

Electron in the Einstein-Weyl Space

S. C. Tiwari
Institute of Natural Philosophy
C/o 1 Kusum Kutir, Mahamanapuri
Varanasi 221005, India

Abstract

The classical unified theory of Weyl is revisited. The possibility of stable extended electron model in the Einstein-Weyl space is suggested.

5.1 Introduction

The Kaluza-Klein unified theory is well known amongst physicists since the advent of higher (> 4) dimensional theories - especially inspired by super-string theories, however Weyl's was the first unified theory of gravitation and electromagnetism [1]. Weyl's theory was rejected on physical grounds by Einstein, Eddington and Pauli, and Weyl himself abandoned it later in the light of the developments in quantum theory: thus the noncompact group of homothetic transformations in the original gauge group was replaced by the compact circle group of phase transformations. It seems very few mathematicians have been interested in Weyl geometry, see remarks in [2], however Dirac in an attempt to revive this theory in 1973 [3] noted that, "Weyl's theory remains as the outstanding one, unrivalled by its simplicity and beauty". I believe there are good reasons to explore Weyl geometry further; (1) influence of Fichtean philosophy on Weyl's thinking [4] and his idea of continuum that, 'a manifold is continuous if the points are joined together in such a way that it is impossible to single out a point just for itself, but always only together with a vaguely delimited surrounding halo, with a neighbourhood' suggest that the study of Weyl geometry may throw light on some fundamental questions in geometry, and (2) even if one considers quantum theory to be the fundamental physical theory, there is a natural generalization of complex Hilbert space of quantum states incorporating Weyl's vector length holonomy such that the lengths of the complex vectors in the Hilbert space are allowed to change under parallel transport, see [5]. Indeed, as Weyl's quantum principle of 1929 [6] & [7] provided ground work for gauge field theories, it also suggests the generalization from the invariant scale factor of quantum mechanics to spaces with a varying scale factor.

A special case of Weyl's theory, namely Einstein-Weyl manifolds have been extensively studied, see [2] for a recent review. An Einstein-Weyl space is defined to be the Weyl space with an extra condition on the curvature, namely the Einstein-Weyl equation, see Equation(21) below. It is well known that the question of the inertial mass of the electron and the stability of the extended structure could not be addressed in the unified theory of Weyl. In this paper we explore the Einstein-Weyl space to get hints for an electron model in a classical geometrical framework. Since some of the definitions and notations are not very familiar in mathematics literature, we summarize them in the next section. In section 3, the action functional is constructed, and the field equations are derived using the variational principle. Concluding remarks constitute the last section.

5.2 Rudiments of Weyl geometry

In the Riemannian geometry, the relative lengths of two vectors at arbitrary distant points can be compared, therefore for a true infinitesimal geometry it is natural to enlarge the general coordinate transformations postulating gauge transformation. Levi-Civita parallelism was used by Weyl to affect this generalization such that besides a metric

$$ds^2 = g_{\mu\nu} dx^\mu dx^\nu, \tag{5.1}$$

there was a linear ground form $A_\mu dx^\mu$. We confine ourselves to four dimensions, and adopt the notations and sign-index conventions as given by Dirac [3] or Eddington [8] with a slight change that A_μ is used for k_μ. Gauge transformation is defined as

$$ds \to ds' = \lambda ds \tag{5.2}$$

$$A_\mu \to A'_\mu = A_\mu + (\ln \lambda), \mu. \tag{5.3}$$

For ordinary derivatives comma (,) is used, and covariant derivative is denoted by Colon (:). A vector under parallel transport from point x^μ to point $x^\mu + dx^\mu$ gets a length change given by

$$\delta l = l A_\mu dx^\mu. \tag{5.4}$$

Total change in length round a small closed loop is determined by the distance curvature,

$$F_{\mu\nu} = A_{\mu,\nu} - A_{\nu,\mu}. \tag{5.5}$$

Generalized tensors in Weyl space are denoted by $^*T$. Tensors which get multiplied by λ^n under the gauge change (2) are called co-tensors of power n, and in-tensors are gauge-invariant ($n = 0$). It can be seen that $g_{\mu\nu}$ is a co-tensor of power 2 and $\sqrt{-g}$ has power 4. One can define co-covariant or Weyl derivative for any co-tensor. To give an example, let S be a co-scalar of power n then its co-covariant derivative is

$$S_{,^*\mu} = S_{,\mu} - nA_\mu S. \tag{5.6}$$

A gauge-invariant affine connection can be obtained to be as

$$^*\tau^\alpha_{\mu\nu} = \tau^\alpha_{\mu\nu} - g^\alpha_\mu A_\nu - g^\alpha_\nu A_\mu + g_{\mu\nu} A^\alpha. \tag{5.7}$$

Finally we give the expressions for the generalized Ricci tensor and scalar curvature

$$^*R_{\mu\nu} = R_{\mu\nu} - 2F_{\mu\nu} - (A_{\mu,\nu} + A_{\nu,\mu}) - g_{\mu\nu}A^\alpha{}_{:\alpha} - 2A_\mu A_\nu + 2g_{\mu\nu}A_\alpha A^\alpha \tag{5.8}$$

and

$$^*R = R - 6A^\mu{}_{:\mu} + 6A_\mu A^\mu. \tag{5.9}$$

Unlike $R_{\mu\nu}$ which is symmetric in μ, ν in the Riemann space, the corresponding tensor $^*R_{\mu\nu}$ contains an antisymmetric part in the Weyl space. Dirac, in his paper [3], makes it symmetric by adding few terms, while Eddington [8] retains the complete expression given above i.e. Equation (8). It seems the difference between the symmetric and antisymmetric Ricci tensors in Weyl space may be related to the two kinds of affine connection [9]. Recall that the covariant derivatives of covariant and contravariant vectors in the Riemann space are different [8]. Proceeding in the same way, the generalised affine connections obtained from the co-covariant derivatives of the covariant and contravariant vectors are different due to the difference in the Weyl powers. One of them leads to the symmetric Ricci tensor, while the other one gives Equation (8).

5.3 Derivation of Einstein-Weyl equations

Let us return to the main objective of this paper. Following our earlier work [10] we present an action functional demanding that the action must be in-invariant. Let us write the action integral

$$S = \int W\sqrt{-g}d^4x. \tag{5.10}$$

Since $\sqrt{-g}$ is a co-invariant of power 4, W must be a co-scalar of power -4. Noting that $^*R$ is a co-scalar of power -2, one may assume $W = {}^*R^2$; this was the choice of Weyl [1]. However, this does not give the Einstein-Weyl equation. A natural assumption preserving the simplicity of the action for Einstein's field equation in Rieamann space is that W is proportional to $^*R$. Thus introducing a co-scalar χ of power -2, it is proposed that

$$W = \xi^* R + 2C\xi^2. \tag{5.11}$$

Here C is an arbitrary real number. To derive the Euler-Lagrange equations of motion, we perform infinitesimal variations in $g_{\mu\nu}$ and A_μ in the action

$$S_{E-W} = \int (\xi^* R + 2C\xi^2)\sqrt{-g} d^4x \tag{5.12}$$

and set δS_{E-W} equal to zero (remember that the variations vanish at and near the boundary of the space-time region). Straightforward calculations give

$$\xi\delta(R\sqrt{-g}) = [\xi(\frac{1}{2}g_{\mu\nu}R - R_{\mu\nu}) + g^{\mu\nu}\xi^{:\alpha}{}_{\alpha} - \xi^{:\mu:\nu}]\sqrt{-g}\delta g_{\mu\nu}, \tag{5.13}$$

$$\xi\delta(A^\alpha{}_{:\alpha}\sqrt{-g}) = (\xi^{:\mu}A^\nu - \frac{1}{2}\xi_{,\alpha}A^\alpha g_{\mu\nu})\sqrt{-g}\delta g_{\mu\nu} - \xi^{:\mu}\sqrt{-g}\delta A_\mu \tag{5.14}$$

and

$$\xi\delta(A_\alpha A^\alpha\sqrt{-g}) = \xi(-A^\mu A^\nu + \frac{1}{2}A_\alpha A^\alpha g^{\mu\nu})\sqrt{-g}\delta g_{\mu\nu} + 2\xi A^\mu\sqrt{-g}\delta A_\mu. \tag{5.15}$$

Substituting $^*R$ from Equation (9) in Equation (12) and using the variational principle we get the following field equations

$$\begin{aligned} &C\xi^2 g^{\mu\nu} + \xi(\frac{1}{2}g^{\mu\nu}R - R^{\mu\nu}) + g^{\mu\nu}\xi^{:\alpha}{}_{\alpha} - 6\xi^{:\mu}A^\nu \\ &+3\xi_{,\alpha}A^\alpha g^{\mu\nu} - \xi^{:\mu:\nu} - 6\xi A^\mu A^\nu + 3\xi A_\alpha A^\alpha g^{\mu\nu} = 0 \end{aligned} \tag{5.16}$$

and

$$\xi^{:\mu} + 2\xi A^\mu = 0. \tag{5.17}$$

Using Equation (17), we can rewrite Equation (16) in the form

$$R^{\mu\nu} = g^{\mu\nu}\Lambda + 2A^\mu A^\nu + 2A^{\mu:\nu}, \tag{5.18}$$

where

$$\Lambda = C\xi + \frac{1}{2}R + A_\alpha A^\alpha - 2A^\alpha{}_{:\alpha}. \tag{5.19}$$

Variation of ξ in the action, Equation (12), gives

$$\xi = -\frac{^*R}{4C}. \tag{5.20}$$

Equation (20) is not an independent one; contraction of Equation (18) gives Equation (20). Substituting ξ in Equation (18), and rewriting the term $2A^{\mu:\nu}$ in symmetric and skew-symmetric parts, we obtain the Einstein-Weyl equation

$$^*S^{\mu\nu} = \frac{1}{4}{}^*Rg^{\mu\nu}. \tag{5.21}$$

Here ${}^*S^{\mu\nu}$ is the symmetric part of the Ricci tensor

$$ {}^*R^{\mu\nu} = {}^*S^{\mu\nu} - 2F^{\mu\nu}. \tag{5.22} $$

Evidently

$$ F^{\mu\nu} = 0 \tag{5.23} $$

in view of the rearranged form of $2A^{\mu:\nu}$ used in Equation (18)

$$ 2A^{\mu:\nu} = A^{\mu:\nu} + A^{\nu:\mu} - F^{\mu\nu}. \tag{5.24} $$

¿From Equation (23) it is clear that A_μ is locally a gradient. This result is consistent with the contracted Bianchi identity for a Weyl space [11], see also [12]. In general

$$ \left({}^*S^{\mu\nu} - \frac{1}{2}{}^*Rg^{\mu\nu}\right)_{:^*\nu} + F^{\mu\nu}{}_{:^*\nu} = 0. \tag{5.25} $$

In 4 dimensions, the Weyl derivative of $F^{\mu\nu}$ is an ordinary covariant divergence, and for the E-W space use can be made of Equation (21) to reduce the Equation (25) in to the form

$$ F^{\mu\nu}{}_{:\nu} = \frac{1}{4}({}^*R^{:\mu} + 2{}^*RA^\mu). \tag{5.26} $$

Substituting Equation (20) and Equation (23) in Equation (26), we get back Equation (17).

This derivation leads to the Einstein-Weyl equations, but for a restricted class of locally conformal Einstein spaces. Examples are known, see [2], such that $F_{\mu\nu}$, is non-vanishing. E-W Equations (21) do not contain an energy-momentum tensor for the field $F_{\mu\nu}$. Therefore this result, namely that the natural E-W space to be a locally conformal Einstein space, seems significant. Is it possible to have a non-zero $F_{\mu\nu}$ together with the E-W equations ? The Bianchi identity Equation (26) indicates that r.h.s. of this equation can be interpreted as a source current density for the Maxwell field. Thus, Equation (17) can be generalized to

$$ \xi^{:\mu} + 2\xi A^\mu = J^\mu. \tag{5.27} $$

For a consistent derivation of the field equations, the action function, Equation (11) should be changed to

$$ W = \xi^*R + 2\xi^2 + pF^{\mu\nu}F_{\mu\nu}. \tag{5.28} $$

Here p is an arbitrary real number, and C is set equal to 1. Variation in $g_{\mu\nu}$ contributes an additional term $2pE^{\mu\nu}$, with the electromagnetic energy-momentum tensor being

$$ E^{\mu\nu} = \frac{1}{4}g^{\mu\nu}F^{\lambda\sigma}F_{\lambda\sigma} - F^{\mu\lambda}F^{\nu}{}_{\lambda}. \tag{5.29} $$

Varying A_μ, instead of Equation (17) we get

$$ F^{\mu\nu}{}_{:\nu} = \frac{3}{2p}(\xi^{:\mu} + 2\xi A^\mu). \tag{5.30} $$

In order to identify Equation (30) with the Bianchi identity, Equation (26), for the E-W space, we assume $p = -3/2$. The co-scalar field ξ determines the current density 4-vector, and Equation (27) can be used to arrive at the following equations:

$$\xi^2 g^{\mu\nu} + \frac{1}{2}\xi^* R g^{\mu\nu} - \xi^* S^{\mu\nu} + B1 + B2 = 0, \tag{5.31}$$

$$B1 = \xi F^{\mu\nu} - \frac{1}{2}[J^{\mu:\nu} - J^{\nu::\mu}] - 4[J^\mu A^\nu - J^\nu A^\mu] \tag{5.32}$$

and

$$B2 = -3E^{\mu\nu} - \frac{1}{2}[J^{\mu:\nu} + J^{\nu:\mu}] - 2[J^\mu A^\nu + J^\nu A^\mu] + J_\alpha A^\alpha g^{\mu\nu}. \tag{5.33}$$

Since $B1$ and $B2$ are trace-less, contraction of Equation (31) gives Equation (20) with $C = 1$. The skew-symmetric part, $B1$, must vanish. It is further postulated that the symmetric part $B2$ is also zero. Equation (31) then reduces to the E-W Equation (21).

At first, assuming $B1$ and $B2$ to be zero may look arbitrary, but returning to the fundamental unresolved problem of the stability of a purely electromagnetic model of electron we gain a new insight for analysing Equation (32) and Equation (33). To simplify the problem, let us consider the E-W space for $\xi =$ a constant. The current continuity equation gives the Lorentz gauge condition as

$$A^\mu{}_{:\mu} = 0, \tag{5.34}$$

and Equation (20) becomes

$$R + 6A_\mu A^\mu = \text{constant}. \tag{5.35}$$

Since $J^\mu = 2\xi A^\mu$, $B1$ is identically zero. Equation (33) shows that for $B2$ to be zero, the electromagnetic energy momentum tensor must balance with a stress tensor of charge density in which the constant ξ appears. The implication that $B2 = 0$ is reminiscent of the Poincare stress for an electron model [13] and the problem of inertial mass and electromagnetic energy in a Weyl space discussed by Weyl [1] and Eddington [8].

5.4 Conclusion

A natural Einstein-Weyl space having pure gauge fields A_μ is shown to follow from a simple action principle. A non-vanishing distance curvature (or the Maxwell field) and the contracted Bianchi identity in the Weyl space indicate a non-trivial generalisation postulating current-density four-vector. The general field Equation (31), is shown to admit a special class of space, namely the E-W space if certain conditions (i.e. $B1 = B2 = 0$) are imposed. Physical arguments emanating from the stability of the extended model of the electron seem to justify these conditions. However, construction of an actual physical model of the electron remains as an exciting possibility. We conclude this paper by asking: "Is the interior of electron an E-W space?"

The library facility of the Banaras Hindu University is acknowledged. I am grateful to Prof. H. Pedersen and Prof. P. Goudochon for useful discussion on the E-W spaces. I thank the anonymous reviewers for constructive comments.

Bibliography

[1] H. Weyl, *Space-time matter*, translated by H.L. Brose (Dover, NY 1952)

[2] D.M.J. Calderbank and H. Pedersen, 'Einstein-Weyl Geometry' in *Essays on Einstein Manifolds* (1998)

[3] P.A.M. Dirac, *Proc. R. Soc.* (London) A333 (1973) 403

[4] E. Scholz, in *Proc. International Congress of Mathematicians* p. 1592 (Birkhauser, 1995)

[5] S.C. Tiwari, 'A note on Weyl-Kahler space' *SCAN 9511230* (CERN); S.C. Tiwari, 'Geometry of quantum theory: Weyl-Kahler space' in *Geometry, Analysis and Applications* edited by R.S. Pathak (World Scientific 2001) p129

[6] H. Weyl, *Zeit. F. Physik* 56 (1929) 330

[7] L. O'Raifeartaigh and N. Straumann, *Rev. Mod. Phys.* 72(2000)1.

[8] A.S. Eddington, *The Mathematical theory of relativity*, 2nd edition (C.U.P. 1924)

[9] I thank one of the reviewers for pointing out this, and drawing my attention to the reference M. Israelit, *The Weyl-Dirac Theory and our Universe*, Nova Science Publisher, Inc (1999)

[10] S.C. Tiwari, *Phys. Lett* A 135 (1989) 315; Phys. Lett. A 142 (1989) 460.

[11] P.G. Bergmann, *Introduction to the Theory of Relativity* (Prentice Hall, 1942)

[12] H. Pedersen and K.P. Tod, *Advances in Math.* 97 (1993) 74.

[13] S.C. Tiwari, *Rebirth of the electron: electromagnetism* (Monograph published by the author 1997)

In: Relativity, Gravitation, Cosmology
Editor: V. Dvoeglazov and A. Espinoza Garrido, pp. 71-88
ISBN 1-59033-981-9

Chapter 6

Toward the Unification of Gravity and Electromagnetism

R. L. Amoroso and J.-P. Vigier
Noetic Advanced Studies Institute – Physics Lab
120 Village Square MS 49
Orinda, CA 94563-2502 USA
E-mail: noeticj@mindspring.com
and
Université Paris VI - CNRS
Gravitation et Cosmologie Relativistes
Tour 22-12 4 ème étage - Boîte 142
4, place Jussieu, 75252 Paris Cedex 05

Abstract

An attempt to unify gravity and electromagnetism in terms of the covariant density distribution of a real average covariant Dirac vacuum built with extended random elements filling flat space-time is presented.

6.1 Introduction

Although the Newton and Coulomb potentials have similar forms the two theories have developed separately leaving their unification an unsolved problem throughout the history of Modern Science. In the past most attempts at unification have been within a frame associating electromagnetism with new geometrical properties of spacetime [1,2,3]. The aim of this paper is different. Following

Puthoff [4], and others, both fields are represented by four-vector field densities A_μ; where one considers both types of phenomena as different types of motions within the same real physical zero-point field in a flat spacetime, i.e. as two different 'vacuum' types of collective perturbations carried by a single 'vacuum' field moving in such a space. Our hope is that since this approach suggests new types of experimentation and new interpretations of unexplained effects it could, if confirmed, help to disentangle the present theoretical discussion.

The basis of this model is as follows:

A) The first basis is observational. The universe apparently does not change with distance [15] (as it would for big-bang type theories). This leads to the possibility of a non-Doppler redshift [5] (which suggests a non-zero photon mass $m_\gamma \neq 0$) with the velocity of light isotropic in an absolute inertial frame I_0, in time.

B) The second basis is that our essential instrument for distance observation (i.e. electromagnetic waves) is more complex than initially thought. De Broglie and Einstein demonstrated that $E = h\nu = \mathrm{mc}^2$ (with $\mathrm{m} = \mathrm{m}_0\left(1 - v^2/c^2\right)^{-1/2}$); now commonly known as the Einstein de Broglie wave equation, where v is velocity compared to the speed of light c. This equation represents the starting point for deriving putative non-zero photon restmass so that individual massive photons could be considered as piloted by real non zero-mass Maxwellian waves allowing the electromagnetic field to be represented by a vector density A_μ. As shown by the Aharonov-Bohm effect [26], this implies that the electromagnetic field is not completely represented by the $\mu\nu$ fields [6,7].

Maxwell's equations [3] conventionally describe Transverse elements denoted as 'EM' waves; by utilizing the Einstein / de Broglie relation one may derive additional degrees of freedom such that Maxwell's equations are not 'cut off' at the vacuum, but lead to Longitudinal wave components and non-zero electric conductivity of the vacuum. Thus our distinct need for the utility of the $\mu\nu$ fields instead of just the standard 'EM'. This also suggests that the photon is 'piloted'. One must 'fix' the coordinates of either the EM field or the $\mu\nu$ field we have chosen the latter. It should be noted that while c is constant in the rest frame and the velocity of massive photons would be frequency dependent ; there is no contradiction because as Dirac himself stated according to coordinate law the pilot wave and the photon decouples [17]. The two sets of coordinates, EM or $\mu\nu$ are exclusive and would generally be independent. It is the aim of this paper to reveal a method for integrating them.

It is well known that the usual form of Maxwell's equations in vacuum (describing zero mass photons) possess infinite families of boundary free exact solutions with Longitudinal electric or magnetic fields; this is the usual B^3 theory where $B^{(6.3)} = 0$ and photon mass $= 0$. This is also true for the vector potential in the Lorentz gauge according to the equation $A_\mu = 0$. But of interest to the task here, for massive photons there is only one family and one set of boundary conditions!

C) The third basis has its theoretical origin in the introduction by Dirac et al. of a real covariant chaotic physical 'aether' which fills space-time, carries real physical observable wave-like and particle like (soliton-like) perturbations or local extended elements, whose four momenta and angular momenta are statistically and evenly distributed on specific hyperbolic surfaces, at each given point, in all given inertial frames. This 'vacuum' distribution thus appears, as invariant isotropic chaotic and undetectable (except in specific physical cases) for all inertial observers. The form taken by an aether within Relativity Theory carrying both particles and waves is now described in terms of collective motions on the top of a real essentially stochastic covariant background. Such an 'aether' theoretically justifies the statistical productions of Quantum Mechanics (in its causal stochastic interpretation) and SED theory, and has a direct experimental justification in the Casimir effect. This implies a background friction (associated with absolute local conservation of total momentum and angular momentum) and collective motions which provide a new interpretation of the observed cosmological red-shift [5, 22] and yields new possibilities to interpret (also in terms of local frictions) the anomalous red-shifts observed by Arp, Tifft and other astronomers [8].

From these bases, section 3, describes the gravitational results of General Relativity in Maxwellian terms. Section 4 develops a possible unification model of both theories. Section 5 is a brief discusses possible consequences of the preceding attempt. This aether is locally defined by a particular real Poincaré frame I_O, in which (measured with real physical instruments) the velocity of light is identical in all directions at all observable frequencies. All observers tied to other frames passing through local inertial motions will see (measure) different spacetime properties (associated with their velocity and orientations) defined by the corresponding Poincaré transformations.[1] Local variations of physical properties of the aether correspond to local transitions relating differential inertial frames at neighboring points.

6.2 Flat Spacetime and a Real Physical Aether

This model depends on the existence of a real physical vacuum (or zero point field) built with extended wave-like individual elements [9, 10] centered on points in an external flat space-time, where such elements can overlap and interact (i.e. carry) collective motions corresponding to excess (electromagnetic 'bumps') or defects (gravitational 'holes') in the average density of the local aether elements. The model could be described as a gas of extended elements within flat spacetime. These elements can interact locally (i.e. carry collective motions) and the gas' local scalar density thus carries waves (and solitons) associated with excess (electromagnetic) or defects (gravitational) in density, with respect to the average local vacuum density. One thus defines field variables associated with these two possible (excess or defect) local density variations. The vector

[1] To quote Kholmetsky "In order to pass from one arbitrary inertial frame I_1 to another one I_2 it is necessary to carry out the transformation from I_1 to the absolute frames I_O and then from I_0 to I_2."

fields, for example, in this paper, represent localized excess or density defects w.r.t. the local vacuum density. This model thus implies:

a) A description of real physical vacuum properties in terms of real extended vacuum elements average behavior.

b) A description of the behavior of its collective excess (above average) associated with recently observed electromagnetic effects.

c) A description of the behavior of its collective defects (below average) associated with observed gravitational effects.

Introducing these new concepts into Maxwell's equations and the description of gravitational fields along the same lines (in terms of vector fields A_μ) suggests a new type of unification of both theories. Instead of looking for a common geometrization of gravity and light (i.e. their unification within a unique form of extended space-time geometry) one could assume the following from Newton and Lorentz:

A) The evolution of extended (fields) and of localized (sources) in terms of 1) vacuum (aether) 2) gravitational fields, 3) the electromagnetic field, reflects the time evolution (motions) and interactions of perturbations of a real material substance moving in a 3-dimensional flat space. This means that all three field and particle sub-elements are localized at given points, at each instant, in this 3-space and move continuously (i.e. locally transform) according to causal laws.[2]

This assumption (distinction of space and fields) is now supported by the existence of a special particular experimental inertial cosmological frame I_0 in which

- The 2.7 $^\circ$K microwave radiation frame is isotropic and non rotating
- The average distribution of different types of galaxies (spiral, elliptical, Q.S.O's) is isotropic and does not change with distance [15].
- The observable anisotropy of the velocity of light propagation in different directions and around massive objects reflects the real motions of real fields described w.r.t. the I_0 frame in any real inertial Poincaré frame by covariant (local) four-vector scalar chaotic average density $\rho(x_\mu)$ around each absolute space-time point x_μ in I_0 i.e. by average four-vectors $A^0_\mu(x_\alpha)$ where ^{0} denotes average measures taken in I_0.[3]

B) That all real physical observations rest on:

1. The utilization of real physical apparatus based on electromagnetic fields and gravitational material with charged (or uncharged) particles.

---

[2] As a consequence of the failure of the geometrical unification program Einstein was still obliged in 1954 to consider the electromagnetic field as filling curved space-time, but never reached a final satisfying model.

[3] This implies 1) the existence of a basic high density of sub-elements in vacuum, 2) the existence of small density variations above (for light) and below (for gravity) the average density with the possibility of propagating density variation on the top of such a vacuum model as initially suggested by Dirac.

2. On observers also built with the same material i.e. influenced by the said fields and particles in Minkowski space.

In other terms all observers (and their observations, inertial or not) are an integral part of fields and particles since they are part of the same overall real field and particle distribution. This fact determines their relation with all real phenomena. A physical theory should explicitly provide (within its context) a definition of the means whereby the quantities with which the theory is built and can be measured. The properties of light rays and massive particles are thus sufficient to provide the means of making basic measurements. Since real clocks and rods are the real instruments utilized in physics, we shall thus first define, for an individual inertial observer, the behavior of such instruments with respect to each other: since this determines, for every inertial observer possessing them, the behavior, with respect to I_0, of the material fields around him.

As a consequence of the covariant distribution character observed in I_0, the very small resistance to motion and assumed non-zero photon rest mass, real spin of possible extended vacuum sub-elements and their internal possible motions (and associated local interactions) one can describe the four-momenta and angular momenta of all extended subelements passing through a small four-volume with a constant average density on a hyperboloid $\sum_0$. The four-momenta and angular momenta of extended elements are distributed at each point$P(x_\mu)$ with constant density $\rho(x_\mu)$ on space-like hyperboloids.

C) Following an idea of Noether the local analysis of moving fields and extended particles at each point by real observers tied to this point, is defined by local clocks and rods which move with the corresponding element. It is thus locally performed at each point of coordinates $x_\mu(\tau)$ which follows a world-line L. To this point are attached local (in I_0) 'internal' variables $b^{(\lambda)}$, which describe its neighborhoods physical properties and thus depend on τ. The evolution is given by $x_\mu(\dot{x}_\mu)$, $b^\lambda(\dot{b}^\lambda)$,where $\cdot$ denotes the proper time derivative w.r.t. τ when x_μ describes a world-line L. A scalar Lagrangian thus represents the evolution of the real physical medium in I_0, which depends on a local Lagrangian L and is thus given by Poisson brackets. This description on I_0 is assumed to correspond to local space-time translations and four dimensional rotations which are determined by a Lagrangian L invariant under the local group of Poincaré transformations (i.e. the inhomogeneous Lorentz group). They contain [15]:

1) The operators P_μ of infinitesimal translations of X_μ only and can be described by $P_\mu \cdot X_\lambda = g_{\mu\lambda}$.

2) The operators $M_{\mu\nu}$ of infinitesimal four rotations in I_0 which act simultaneously on X_μ and on the internal variables. We have at X_μ:

$$M_{\mu\nu}x_\lambda = x_\mu g_{\nu\lambda} - x_\nu g_{\mu\lambda}. \tag{6.1}$$

Their action on internal local variables depends on their choice.

3) A choice of L leads to the momenta

$$G_\mu = \frac{\partial L}{\partial \dot{x}_\mu} \quad \text{and}\beta^{(\lambda)} = \frac{\partial L}{\partial \dot{b}^{(\lambda)}} \tag{6.2}$$

yielding a constant impulsion vector

$$G_\lambda P_\mu x_\lambda = G_\lambda g_{\mu\lambda} = G_\mu \quad : \tag{6.3}$$

and the total angular momentum:

$$M_{\mu\nu} = G_\lambda M_{\mu\nu} x_\lambda + \beta^{(\lambda)} M_{\mu\nu} b^{(\lambda)},$$

so that $M_{\mu\nu} = x_\mu G_\nu - x_\nu G_\mu + S_{\mu\nu}$, with $S_{\mu\nu} = \beta^{(\lambda)} M_{\mu\nu} \beta^{(\lambda)}$.

These quantities satisfy the Inhomogeneous Lorentz group commutation relations $[P_\mu, P_\lambda] = 0$

$$[M_{\mu\nu}, P_\alpha] = g_{\alpha\beta} P_\nu - g_{\alpha\nu} P_\mu \tag{6.4}$$

i.e. Poisson Group Relations:

$$[G_\mu, G_\nu] = 0 \quad [M_{\mu\nu}, G_\alpha] = g_{\alpha\beta} G_\nu - g_{\alpha\nu} G_\mu \tag{6.5}$$

$$[M_{\mu\nu}, M_{\alpha\beta}] = g_{\mu\alpha} M_{\nu\beta} + g_{\nu\beta} M_{\mu\alpha} - g_{\mu\beta} M_{\nu\alpha} - g_{\nu\alpha} M_{\mu\beta}.$$

With these quantities one can also define local conservation laws for 'free' elements i.e.

$$\begin{aligned} \dot{G}_\mu &= 0 \\ \dot{M}_{\mu\nu} &= 0 \\ \dot{S}_{\mu\nu} &= G_\mu \dot{x}_\nu - G_\mu \dot{x}_\mu. \end{aligned} \tag{6.6}$$

and introduce a constant local mass term M_0 with $G_\mu G_\mu = -M_0^2 \cdot c^2$. The mass term M_0 is the associated local inertial mass generated by the rotating pairs of subelements (as in Fig. 1). M_0 is located Y_μ (Eq. 9) the center of gravity of the rotating subelements in I_0.

4) An associated center of gravity y_μ is defined by the introduction of the four-vector

$$R_\mu = \left(\frac{1}{(M_0^2 c^2)}\right) \cdot S_{\mu\nu} \cdot G_\nu \tag{6.7}$$

associated with x_μi.e. $y_\mu = x_\mu - R_\mu$; (9) which implies that locally extended real media in I_0 are described by pairs of points as first suggested by Yukawa.

5) An inertial mass (usually not constant) μ_0 (relating to the inertial frame in I_0, while M_0 is a new mass term centered on a point as illustrated for the 'bump' and 'hole' configuration in Figure 1) is defined by

$$-M_0 c^2 = G_\mu \cdot \dot{x}_\mu \tag{6.8}$$

can also be attributed to x_μ: M_0 being located at y_μ (Fig. 1) since one has: $\dot{y}_\mu = \dot{x}_\mu - \dot{R}_\mu = \dot{x}_\mu - \frac{1}{M_0^2 c^2}(G_\mu \cdot \dot{x}_\nu - G_\nu \dot{x}_\mu) G_v = \frac{\mu_0}{M_0^2} \cdot G_\mu$ so that the motion of $y_\mu$is locally rectilinear and y_μ has a proper time Θ, (with $d\lambda/d\Theta = M_0/\mu_0'$) and we have: $y_\mu' = \dot{y}_\mu \cdot \frac{d\tau}{d\Theta} = G_\mu / M_0$ =constant. and

$$\mu_{\mu\nu} = R_\mu G_\nu - R_\nu G_\mu + S_{\mu\nu},$$

w.r.t. the center of gravity. Local instantaneous four rotations are described by:

- A specific 'beigrössen' four-frame b^{ξ}_{μ} (ξ=1,2,3,0) with $\dot{x}_{\mu} = b^4_{\mu} = \frac{ic}{6}\varepsilon_{\mu\nu\alpha\beta} \cdot \varepsilon^{rst} b^r_{\nu} b^s_{\alpha} b^t_{\beta}$, $b^{\xi}_{\mu} = (i/2)\varepsilon_{\mu\nu\alpha\beta}\dot{x}_{\nu} S_{\alpha\beta}$ and $S_{\alpha\beta} = I \cdot \dot{b}^{\xi}_{\alpha} \cdot b^{\xi}_{\beta}$.

- A specific four-frame a^{ξ}_{μ} centered on y_{μ} with $M_{\alpha\beta} = K \cdot \dot{a}^{\xi}_{\alpha} \cdot a^{\xi}_{\beta}$ for a^4_{μ}

along $y'_{\mu}$and $a^3_{\mu} = (i/2M_0 c) \cdot \varepsilon_{\mu\nu\alpha\beta} G_{\nu} \mu_{\alpha\beta}$.

This set of relations must be completed by relations which will define the interactions between the extended elements i.e. the propagation in the aether of collective motions corresponding to observed gravitational and electromagnetic phenomena.

Before the introduction of such interactions one must recall that such proposals have already been made in the past. We only mention here:

- Weyssenhof's proposal [9] $S_{\alpha\beta}\dot{x}_{\beta} = 0$ extensively discussed in the literature.
- Nakano's proposal [12] $S_{\alpha\beta}\dot{x}_{\beta} = I \cdot \ddot{x}_{\alpha}$.
- Roscoe's proposal with photon mass [13].

6.3 General Relativity Represented as a Polarizable Vacuum

Since all observed effects of gravity in distant space rest on light observation (including γ and radio *EM* waves coming through space from distant sources) a simple model endows the polarizable vacuum with properties that might account for all the phenomena in terms of distortions. This initial proposal of Wilson [24] and Dicke [25] has been recently revived with astonishing success by Puthoff [4] and Andrews [4] and Krogh [14]. We first summarize their model and will complete it with a supplementary mass term in electro-magnetism. Our initial purpose here is to demonstrate the possible 'geometric' relationship between G and EM in terms of the Dirac Vacuum.

One starts from the idea that in flat space the electric field moves in a real "vacuum medium" with a point varying dielectric constant K: so that this D field satisfies the vacuum equation:

$$D = K \cdot \varepsilon_0 \cdot E. \tag{6.9}$$

This corresponds to a variable fine structure constant

$$\alpha = \frac{e^2}{4\pi\varepsilon_0 \hbar c} \cdot \left(\frac{\mu(K)/\mu_0}{K}\right)^{1/2} : \tag{6.10}$$

so that the vacuum has permittivity and permeability constants given by $\varepsilon_0 \rightarrow \varepsilon = K \cdot \varepsilon_0$ and $\mu_0 \rightarrow \mu = K \cdot \mu_0$, and an impedance $(\mu/\varepsilon)^{1/2} = (\mu_0/\varepsilon_0)^{1/2}$ to satisfy Eötvos-type experiments. The local velocity of light for a given frequency ν varies like $V_{\nu} = c/K$ i.e. like $1/(\mu\varepsilon)^{1/2}$. The corresponding principle of equivalence implies that the self-energy of a system changes when K changes; so that a flat-space energy E_0 in flat space changes into

$$E = E_0 \cdot (K)^{-1/2}; \tag{6.11}$$

and one has $m = m_0 \cdot K^{3/2}$.

As a consequence the condition E $=\hbar \cdot \omega$ becomes

$$\omega = \omega_0 (K)^{-1/2} \tag{6.12}$$

along with the time and length variations Δt and Δr given by the relations: $\Delta t = \Delta t_0 (K)^{1/2}$ and $\Delta r = \Delta r_0 (K)^{-1/2}$.

These relations are evidently equivalent to a local curvature of space. Indeed a dx_0 length rod shrinks to $d_x = d_{x_0} \cdot (K)^{-1/2}$and would measure dx_0, where the rod remains rigid, is now expressed in terms of dx-length rod as $dx_0 = (K)^{1/2} dx$.

Using the same argument for dt and dt_0 we find that one can write:

$$dS^2 = c^2 dt_0^2 - (dx_0^2 + dy_0^2 + dz_0^2) \tag{6.13}$$

which transforms into

$dS^2 = \frac{1}{K} c^2 dt^2 - K(dx^2 + dy^2 + dz^2)$: (a) i.e. $dS^2 = g_{ij} ... dx^i dx^j$, (b)

with $g_{00} = 1/K, g_{11} = g_{22} = g_{33} = -K$ and $g_{ij} = 0$ for $i \neq j$.

In the case of a spherically symmetric mass distribution one writes

$$\begin{cases} K = e^{2G \cdot M / rc^2} \\ K = 1 + 2\frac{G \cdot M}{rc^2} + \frac{1}{2}\left(\frac{2GM}{rc^2}\right)^2 + ... \end{cases} \tag{6.14}$$

where G is the gravitational constant, M the mass and r the distance from its origin located at the center of mass. Puthoff [5] has recently shown that this model accounts (sometimes with better precision) for all known experimental tests of General Relativity in a simple way i.e. one can describe

- The gravitational redshift given by $\omega = \omega_0 / (K)^{1/2}$ (so that $\cong (\Delta\omega / \omega GM / R^2 c^2) h$ has a 1/100 precision).
- The bending of light rays by the sun and stars.
- The advance of the Perihelion of Mercury.

He has also shown that one can derive the form of (6.14) from a general Lagrangian with a variable K i.e. leaving aside vacuum interaction,

$$\begin{aligned} L = &- \left[\frac{m_0 c^2}{K^{1/2}} \left(1 - \left(\frac{\nu}{(c/K)} \right)^2 \right)^{1/2} + q \cdot \phi - q \cdot \vec{A}\vec{V} \right] \delta^3 (r - \bar{r}) \\ &- \frac{1}{2} \left(B^2 / (K \cdot \mu_0) - K(\varepsilon_0 E^2) \right) - \frac{\lambda}{K^2} \left[(\nabla K)^2 - \frac{1}{(c/K^2)} \left(\frac{\partial K}{\partial t} \right)^2 \right] \end{aligned} \tag{6.15}$$

in I_0.

This association of gravitational theory with electromagnetic theory based on the introduction of a variable dielectric 'vacuum' constant K has recently been made more explicit by Krogh [14]. Noting that:

a) Electromagnetic theory implies the effects of electromagnetic vector four-potential vectors A_μ on the phases S of quantum mechanical waves so that one has

$$\Delta S = \frac{q}{h}\int \phi dt - \frac{q}{hc}\int \vec{A}\cdot d\vec{S} \tag{6.16}$$

for charged particles moving under the influence of the four vector, A_μ.

b) If $m_\gamma \neq 0$ ($m_\gamma$is the mass term introduced into Maxwell's equation) the force on charged particles takes the form $F = q\left(E + \frac{V\times B}{c}\right) + q\cdot V$ (where the first term is the usual transverse Poynting force on currents and the second a longitudinal force along currents (resulting from non zero photon mass) recently observed by Graneau [11] and Saumont [16].

c) One can describe gravity with a four-vector density A^g_μ so that the gravitational (Newton) and electromagnetic (Coulomb) potentials have the same form, but different coupling constants. This suggests that both wave fields and singularities are just different aspects of the same fundamental field.

6.4 Maxwell's Equations Extended

This discussion opens the possibility to test new types of extensions of Maxwell's equations in the laboratory. Since this has already been attempted some results (derived within the frame of the model) are given here:

a) From a non-zero vacuum conductivity coefficient $\sigma \neq 0$[6] we have in vacuum div E =0 with curl $H = \sigma E + \varepsilon_0\chi_0\partial E/\partial t$ and div $H = 0$ with curl $E \quad = -\mu_0\chi_m\partial H/\partial t$.

b) From an associated non-zero photon mass term ($m_\gamma \neq 0$) (with $A_\mu A_\mu \to 0$ whereA_μ denotes the total four-potential density in Dirac's aether model. This introduces a non-zero fourth component of the current $J_\mu = \sigma E, j_0$ (where $j_0 \neq 0$)into the vacuum corresponding to a real detectable space. With present technology this implies that the present 'vacuum' really carries space-charge currents [17] (so that the divergence of the electric field is different from zero 'in Vacuo') and the corresponding existence of a displacement current (i.e. a curl of the magnetic field) and its associated current density[4].

6.4.1 The Infinitesimal Mass of Photons

Unifying massive spin 1 photons piloted by electromagnetic waves built with massive extended sub-elements has been developed in a series of books by Evans, Vigier et al. [6] The model implies the introduction of spin and mass with an

---

[4]Such attempts have been recently published in a book by Lehnert & Roy [18] so we shall only present a summary of some results and assumptions.

associated energyless magnetic field component $B^{(3)}$ in the direction of propagation and a small electrical conductivity in the Dirac vacuum also implying a new 'tired light' mechanism [6, 22]. Corresponding equations will be given below.

In the " absolute " inertial frame I_0 all massive particles are governed by a gravitational potential four-vector $\phi_g, \vec{A}_g/c$, associated with a small mass m_g which can be decomposed into transverse, longitudinal and gradient potentials.

We can thus associate the relations

$\sim \phi = -\frac{\rho}{\varepsilon_0} + \mu\phi\text{gnd} \sim \quad \vec{A} = -\vec{d}_0/\varepsilon_0 c + \mu\vec{A}$ which represent the electromagnetic field in vacuum in any inertial frame $\sum_0 therelations$:

$\sim \phi_g = 4\pi G m \rho_\mu + \mu_g \cdot \phi_g$ and $\sim \vec{A}_g = 4\pi \cdot G \cdot \vec{\jmath}_m + \mu_g \vec{A}_g$, which represent the gravitational field in the same vacuum; where ρ_μ refers to mass density, $\jmath_m$ to mass current and μ and μ_g to electromagnetic and gravitational mass (both very small $\cong 10^{-65}$ grams) and $\rho \cdot c_0$ in the $\sim$ terms ($\sim$ g $\nabla^2 - (1/c_0^2)\partial/\partial t^2$) represents the corresponding wave velocities (which except in I_0 depend on the directions in flat space-time) so that one has:

$$c_0 = c \cdot e^{2\phi_g/c^2}; \tag{6.17}$$

where c is the value in the absence of a gravitational potential A_μ^g. In this model, one assumes, with Sakharov, that the gravitational field corresponds to local depressions in the immensely positive energy of the zero-point field; and gravitational fields represent regions of diminished energy (i.e. that their momentum gravity corresponds to " holes " in vacuum energy or local defects of vacuum elements). Their effective momentum is thus opposite and corresponding gravitational forces are attractive.

Such an association also suggests that although measuring devices (observations) in local inertial Poincaré frames are altered by gravitational potentials (they are part of the same real physical background in this model). There is no effect on the geometry of flat space and time. For any given real inertial local Poincaré frame Σ_0 real space is Euclidean and one uses Poincaré transformations between Σ_0 and I_0 to describe real motions which include consequences of gravitational potentials. For example a reduction of the velocity of quantum mechanical waves, including light, is taken as a fundamental effect of gravitational potentials. Clocks are slowed and measuring rods shrink in such potentials by a factor e^{ϕ_g/c^2}.

6.4.2 Divergence of the Electromagnetic Field

A non-vanishing divergence of the electric field given below, can be added to Maxwell's equations which results in space-charge distribution. A current density arises in vacuo and longitudinal electric non-transverse electromagnetic terms (i.e. magnetic field components) appears (like $B^{(3)}$) in the direction of propagation.

Both sets of assumptions were anticipated by de Broglie and Dirac. They imply that the real zero-point (vacuum) electromagnetic distribution

- is not completely defined by $F_{\mu\nu}$ but by a four-vector field distribution given by a four-vector density A_μ associated with a de Broglie-Proca equation i.e. $\sim A_\mu(x_\alpha) = -\frac{m_\gamma^2 c^2}{\hbar^2} A_\mu(x_\alpha)$ and its complex conjugated equation.
- the A_μ field potential equation also contains a gradient term so one has in vacuum (6.13):

$$A_\mu = A_\mu^T + A_\mu^L + \lambda \partial_\mu S \tag{6.18}$$

where S is the scalar for space with $A_\mu A^* \to 0$ and a small electrical conductivity in vacuo.

6.5 Some Possible New Consequences of the Theory

Since such models evidently imply new testable properties of electromagnetic and gravitational phenomena we shall conclude this work with a brief discussion of the points where it differs from the usual interpretations and implies new possible experimental tests.

If one considers gravitational and electromagnetic phenomena as reflecting different behaviors of the same real physical field i.e. as different collective behavior, propagating within a real medium (the 'aether') one must start with a description of some of its properties.

We thus assume:

A) that this 'aether' is built (i.e. describable) by a chaotic distribution $\rho(x_\mu)$ of small extended structures represented by four-vectors $A_\mu(x_\alpha)$ round each absolute point in I_0. This implies

- The existence of a basic local high density of extended sub-elements in vacuum
- The existence of small density variations $\delta\rho(x_\mu)A_\alpha(x\mu)$ above $\delta\rho > 0$ for light and below $(\delta\rho < 0)$ for gravity density at x_μ.
- the possibility to propagate such field variations within the vacuum as first suggested by Dirac [17].

One can have internal variations: i.e. motions within these sub-elements characterized by internal motions associated with the internal behavior of average points (i.e. internal center of mass, centers of charge, internal rotations : and external motions associated with the stochastic behavior, within the 'aether', of individual sub-elements. As well known the latter can be analyzed at each point in terms of average drift and osmotic motions and A_μ distribution. It implies the introduction of non-linear terms. Tysis has been developed by MacGregor [4], Guerra and Pusterla and Smolin.

To describe individual non-dispersive sub-elements within I_0, where the scalar density is locally constant and the average A_μ equal to zero, one introduces at its central point $Y_\mu(\theta)$ a space-like radial four-vector $A_\mu = r_\mu \exp(iS/\hbar)$ (with $r_\mu r^\mu = \mathrm{a}^2 = \text{constant}$) which rotates around Y_μ with a frequency $\nu = m_\gamma c^2/h$. At both extremities of a diameter we shall locate two opposite electric charges

e^+ and e^- (so that the subelement behaves like a dipole). The opposite charges attract and rotate around Y_μ with a velocity $\cong c$. The +e and −e electromagnetic pointlike charges correspond to opposite rotations (i.e. $\pm\ \hbar/2$) and A_μ rotates around an axis perpendicular to A_μ located at Y_μ, and parallel to the individual sub-element's four momentum $\partial_\mu S$.

Assuming electric charge distributions correspond to δm >0 and gravitation to $\delta m < 0$ one can describe F.A.P.P. such sub-elements as holes ($\delta m < 0$) around a point 0 around which rotate two point-like charges rotating in opposite directions as shown in Figure 1 below.

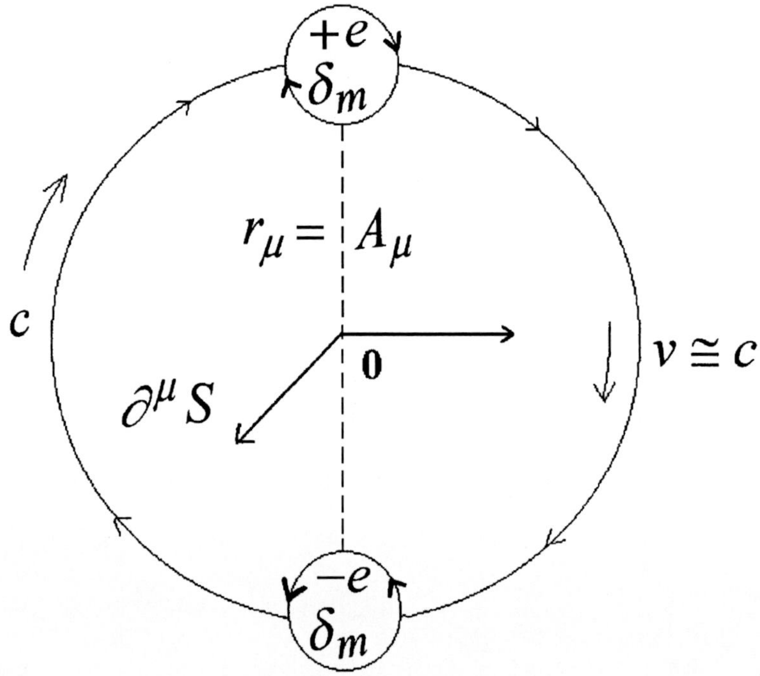

Figure 6.1: Diagram conceptualizing two oppositely charged subelements rotating at v≅ c around a central point 0 behaving like a dipole 'bump' and 'hole' on the topological surface of the polarized Dirac vacuum.

These charges themselves rotate with a velocity c at a distance $r_\mu = A_\mu$(with $r_\mu r_\mu$ = Const.). From 0 one can describe this by the equation

$$\sim A_\mu - \frac{m_\gamma^2 c^2}{\hbar^2} \cdot A_\mu = \frac{[]\,(A_\alpha^* A_\alpha)^{1/2}}{(A_\alpha^* A_\alpha)^{1/2}} \cdot A_\mu \tag{6.19}$$

with $A_\mu = r_\mu \cdot \exp\left[iS(x_\alpha)/\hbar\right]$ along with the orbit equations for e^+ and e^- we get the force equation $m \cdot \omega^2 \cdot r = e^2/4\pi r^2$ and the angular momentum equation:

$$m_\gamma \cdot r^2 \cdot \omega = \hbar/2 \tag{6.20}$$

Eliminating the mass term between (6.19) and (6.20) this yields

$$\hbar\omega = e^2/2r \tag{6.21}$$

where $e^2/2r$ is the electrostatic energy of the rotating pair. We then introduce a soliton-type solution $A^0_\mu = \frac{\sin \cdot K \cdot r}{K \cdot r} \cdot \exp[i(\cot - K_0 x)]$ where $K = mc/\hbar$, $\omega = mc^2/\hbar$ and $K_0 = mv/\hbar$ satisfies the relation (6.19) with $r = ((x-vt)^2 \cdot (1 - v^2/c^2)^{-1} + y^2 + z^2)^{1/2}$ i.e. $\sim A^0_\mu = 0:$ g so that one can add to A^0_μ a linear wave A_μ (satisfying $\sim A_\mu = (m^2_\gamma c^2/\hbar^2)A_\mu$) which describes the new average paths of the extended wave elements and piloted solitons.

Within this model the question of the interactions of a moving body (considered as excess or defect of field density, above or below the 'aether's' neighboring average density) with a real 'aether' appears immediately[5].

As well known, as time went by, observations established the existence of unexplained behavior of light and some new astronomical phenomena which led to discovery of the Theory of Relativity.

In this work we shall follow a different line of interpretation and assume that if one considers particles, and fields, as perturbations within a real medium filling flat space time, then the observed deviations of Newton's law reflect the interactions of the associated perturbations (i.e. observed particles and fields) with the perturbed average background medium in flat space-time. In other terms we shall present the argument (already presented by Ghosh et al. [19]) that the small deviations of Newton's laws reflect all known consequences of General Relativity

The result from real causal interactions between the perturbed local background " aether " and its apparently independent moving collective perturbations imply absolute total local momentum and angular momentum conservation resulting from the preceding description of vacuum elements as extended rigid structures.

6.6 Extending Newton's Model with Inertia and Vacuum Drag

Starting from an " aether " built with moving small extended structures with an average real distribution isotropic in an inertial frame I_0 (i.e. examining the effects in a given inertial frame I centered on a point Y_μ of the real vacuum distribution on a test particle moving with absolute velocity V^0 and angular momentum $\omega^0_{\alpha\beta}$) one can evaluate more precisely, the collective interactions carried by this " aether " between two extended neighboring regions centered on points A and B with two centers of mass situated at X_A and X_B.

[5] According to Newton massive bodies move in the vacuum, with constant directional velocities, i.e. no directional acceleration, without any apparent relative " friction " or " drag " term. This is not true for accelerated forces (the equality of inertial and gravitational masses are a mystery) and apparent absolute motions proposed by Newton were later contested by Mach.

If we start with $\delta\rho < 0$ i.e. gravitational effects, it appears immediately

a) if one assumes the gravitational potential is spherical in the rest frame I_B of its source B,

b) that the motion of A undergoes a velocity dependent inertial induction w.r.t. A i.e. a friction depending on the velocity v of A w.r.t. B

c) that this motion is also submitted to an acceleration dependent inertial w.r.t. I_B i.e. also an acceleration depending on its acceleration a measured in I_B.

d) possible terms depending on higher order time derivations which we will neglect in the present analysis we can write the force on A due to B in I_B in the form $F = F_S + F_v + F_a$ where

$$F = -G \cdot \frac{m_A \cdot m_B}{r^2} - G' \frac{m_A \cdot m_B}{c^2 r^2} \cdot v^2 \cdot f(\theta) \hat{U}_r - G'' \cdot \frac{m_A \cdot m_B}{c^2 r} \cdot a f(\phi) \hat{U}_r \quad (6.22)$$

The terms G, G', G'' are scalars possibly dependent on v. The terms m_A and m_B are the gravitational masses in I_B, $\hat{U}$, is the unit vector along $r.f(\theta)$ and $f(\phi)$ must have the same form i.e. $1/2 \cos\phi$ or $\cos\phi\,|\cos\phi|$. If we also accept the preceding velocity dependent analysis for contracting rods and retarded clocks then we should write $G = G'$ in (6.22) and take f (θ) = =cos $\theta\,|\cos\phi|$as done by Ghosh [19]. Moreover, if we compare the form given by Weber to the repulsion of two electric charges of the same sign : $F^e_{AB} = \frac{e_A \cdot e_B}{4\pi\varepsilon r^2} \cdot \left[1 - \frac{1}{c^2}\left(\frac{dr}{dt}\right)^2 + \frac{2r}{c^2} \cdot \frac{d^2r}{dt^2}\right]$ corresponding to electromagnetism, with the recent form given by Assis [18, 23] to attracting interacting masses m_A and m_B i.e.

$$F^g_{AB} = -G \cdot \frac{m_A . m_B}{r^2} \left[1 + \frac{6}{c^2}\left[r \cdot \frac{d^2r}{dt^2} - \frac{1}{2}\left(\frac{dr}{dt}\right)^2\right]\right] \quad (6.23)$$

We see they have exactly the same form; the difference of their coefficients being compatible (within our interpretation) since they correspond to opposite variations of the average vacuum density. Their interpretation in terms of $\delta\rho > 0$(for electromagnetism) and $\delta\rho < 0$ (for gravitation) also explains (at last qualitatively) why extended depressions repel or attract when they rotate through parallel or antiparallel directions and only attract when they rotate through parallel or antiparallel directions and only attract when $\delta\rho < 0$. This also explains why a reduction of attraction between two masses has been observed when one puts another mass between them (the LAGEOS satellite). In this model this similarity is indeed comparable to similar behaviors of vortices for gravitation and Tsunamis for electromagnetism on an ocean surface.

If one assumes the absolute local conservation of four-momentum and angular momentum in regions containing the preceding " aether " carrying its associated collective electromagnetic and gravitational motions one can evaluate the effects of their interactions. With a real physical " aether " there is no such thing as " free " electromagnetic or gravitational phenomena. Drag theories (described as " inertial induction ") are always present and responsible for Casimir type effects in the microscopic domain. Real consequence of the " aether " appear, at various levels, in the macroscopic and cosmological domains ... as has already been suggested in the literature and tested in laboratory or astronomical phenomena. We only mention here:

1) Possible consequences of modifying and testing the Newton and Coulomb forces.

2) The redshift and variable velocity of electromagnetic waves which results from the rotational inertial drag of extended photons moving in vacuum: an effect already observed in light traversing around the earth [20].

3) The possible measurable existence of the redshift of transverse gravitational waves ... possible in the near future.

4) Observational redshift variations of light emitted by Pioneer close to the solar limb i.e. also of photons grazing a massive object [20].

5) The observed anisotropy of the Hubble constant in various directions in the sky [20] associated with various galactic densities.

6) Observed torques on rotating spheres in the vicinity of large massive bodies. This also appears in some experiments, i.e.:

1. Secular retardation of the earth's rotation.
2. Earth-moon rotation in the solar system etc.

7) Apparent evolution with time of angular momentum in the solar-planetary system.

8) Different variation of redshift of light traveling up and down in the Earth's gravitational field... Which also supports existence of photon mass.

6.7 Summation and Conclusions

This model exploits as stated in the introduction:

a) the analogy (underlined by Puthoff) between the four vector density representation of gravity and electromagnetism in flat space-time [4]

b) the possibility of describing the causality of quantum mechanical phenomena in terms of extended solitons piloted i.e. by quantum mechanical potentials, by real guiding collective waves on a chaotic, polarizable Dirac-type aether - both moving in a flat space-time [20].

c) the representation of this " real vacuum " (Dirac aether) in terms of the chaotic distribution of real extended elements moving in the flat space-time.

d) the introduction of internal motions within extended sub-elements and their relation with local collective motions i.e. the $E = mc^2 = h\nu$ relation

e) the representation of the electron (and its associated pilot-wave) in terms of extended elements with a point-like charge rotating around a center of mass [20].

These assumptions yield realistic physical characteristics to known empirical properties and predict new testable relations besides known properties of elementary particles. The present model must thus be extended, by associating new internal motions to these known properties and interpret them in terms of new strong spin-spin and spin-orbit interactions.

Our attempt is justified by the existence of electromagnetic phenomena not explained by Maxwell's equations. Barrett [21] has stated that Maxwell's theory does not explain the Aharonov-Bohm (AB) effect [26] and Altahuler-Aharonov-Spivak (AAS) effects. It does not cover the topological phase question i.e. the Berry-Aharonov-Anandan, Pancharatnam and Chio-Wu phase-rotation effects. An inclusion of Stoke's theorem is necessary and results of Ehrenberg and Siday must be analyzed. The quantum results of Josephson, Hall, de Haas and van Alphen Sagnac-type experiments also need clarification.

The integration of gravity and electromagnetism however, is not finished, because unification is so far only accomplished in terms of 'bumps' and 'holes' rotating on the stochastic surface of the polarized Dirac Vacuum. Unification must also occur in terms of the richer Higher Dimensional (HD) structure of vacuum topology where one would show the geometric origin of charge and how 'bumps' and 'holes' transform into each other through quasi-particle like transitions 'piloted' by advanced and retarded potentials of the fundamental unitary field itself.

Bibliography

[1] A. Einstein, *Geometry and Experience Sidelights on Relativity*, Denver, 1922.

[2] L. Bass and E. Schrödinger, *Proc. Roy. Soc. A* 232, 1955.

[3] J.C. Maxwell, *A Treatise on Electricity and Magnetism*, New York: Dover (1954).

[4] H.E. Puthoff, 2002, Polarizable-vacuum (PV) approach to General Relativity, *Foundations of Physics*, 32 :6, pp. 927-43; Polarizable-Vacuum, Representation of General Relativity, in R.L. Amoroso , G. Hunter, M. Kafatos & J-P Vigier (eds.), *Gravitation and Cosmology: From the Hubble Radius to the Planck Scale*, 2002 Dordrecht: Kluwer ; Gravity as a zero-point-fluctuation force, *Phys. Rev. A*, 1989, 39 :5, pp. 2333-42 ; T.B. Andrews, *Observed Tests and Theory of the Static Universe*, 2000.

[5] R.L. Amoroso, M. Kafatos, & P. Ecimovic, 1998, The origin of cosmological redshift in spin exchange vacuum compactification and nonzero restmass photon anisotropy, in G. Hunter, S. Jeffers & J-P Vigier (eds.), *Causality & Locality in Modern Physics*, pp. 23-28, Dordrecht: Kluwer.

[6] M.W. Evans and J.P. Vigier, *The Enigmatic Photon*, Kluwer, 1996.

[7] L. de Broglie, Mécanique Ondulatoire du Photon, Paris: Gauthier-Villas (1955). L. de Broglie and J.P. Vigier, *Phys. Rev. Lett.* D. Bohm & J-P Vigier, *Phys. Rev.* 109 (1958), p. 1882.

[8] H. Arp, *Quasar Redshift and Controversies* New York : Interstellar Media, USA, 1987.

[9] J.V. Weyssenhof and A. Raabi, *Act. Phys. Pol.* 9.

[10] J.P. Vigier, *Phys. Lett.* A., 235, 1997.

[11] P. Graneau, 2002, *Why does lightning explode and generate MHD*, preprint.

[12] I. Nakano, *Progress in Theoretical Physics*, (1956), 15.

[13] D.F. Roscoe, *Maxwell's equations*, 1997, preprint.

[14] K. Krogh, Gravitation without Curved Space-Time, 2002, Submitted to *General Relativity & Gravitation.*

[15] F. Halbwachs, J.M. Souriau and J.P. Vigier, *Journal de Physique et de Radium*, 1922, 22, p. 26. See also F. Halbwachs, *Théorie Relativiste. Fluides.* Paris : Gauthier-Villars, 1960.

[16] F. Saumont, Proc. *Cold Fusion and Now Energy Symposium*, Manchester, 1998.

[17] P.A.M. Dirac, 1973, New ideas of space and time, Naturwissenschaften, 32 :6, pp. 529-31 ; *Nature*, 1951, 906.

[18] 18. B. Lehnert, & S. Roy, *Extended Electromagnetic Theory*, Singapore : World Scientific, 2000.

[19] A. Ghosh, *Origin of Inertia*, Apeiron, Montreal, 2000.

[20] R. Pound and G.A. Rebka, *Physical Review Letters* 4, 337, 1960. J-P Vigier & S, Maric, to be published.

[21] T.W. Barrett, Electromagnetic Phenomena not Explained by Maxwell's Equations: *Essays on the Formal Aspects of Electromagnetic Theory.* Singapore: World Scientific, (1993).

[22] R.L. Amoroso, & J.-P. Vigier, The Origin of CMBR as Intrinsic Blackbody Cavity-QED Resonance Inherent in the Dynamics of the Continuous State Topology of the Dirac Vacuum, in R.L. Amoroso, G. Hunter, M. Kafatos & J-P Vigier (eds.), Gravitation and Cosmology: From the *Hubble Radius to the Planck Scale*, 2002 Dordrecht: Kluwer.

[23] Assis, AKT, 1992, Deriving gravitation from electromagnetism, *Canadian J of Physics*, 70 :5, pp. 330-40.

[24] Wilson, H.A., 1921, An electromagnetic theory of gravitation, *Phys. Rev.* 17, pp. 54-59.

[25] Dicke, R.H., 1957, Gravitation without a principle of equivalence, *Rev. Of Mod. Phys.* 29 :3, pp. 3-37.

[26] Aharonov, Y, & Bohm, D. 1959, *Phys. Rev.* 115, 485.

In: Relativity, Gravitation, Cosmology
Editor: V. Dvoeglazov and A. Espinoza Garrido, pp. 89-95
ISBN 1-59033-981-9

Chapter 7

Time Evolution of a Quantum Particle and a Generalized Uncertainty Principle

A. Camacho
Physics Department,
Universidad Autónoma Metropolitana-Iztapalapa.
P. O. Box 55-534, 09340, México, D.F., México
email: acamacho@janaina.uam.mx

Abstract

Some of the possible consequences of a generalized uncertainty principle (which emerges in the context of string theory and quantum gravity models as a consequence of fluctuations of the background metric) are analyzed considering the case of a quantum particle immersed in a homogeneous gravitational field. It will be shown that the expectation value of the momentum operator depends in a novel way on the mass of the involved particle. This kind of physical characteristics could be, in principle, detected. In other words, one way of confronting against the experiment some of the models around quantum gravity is given by the detection of the dependence upon the mass parameter of the expectation value of the momentum operator.

7.1 Introduction

One of the most difficult tasks of modern physics comprises the reconciliation of quantum theory (QT) with general relativity (GR). Though this is still an open problem, there are already some models that appear as feasible candidates. Among them we may mention string theory and quantum gravity, models that entail a modification of the well known uncertainty principle [1, 2]. The new form of this principle is closely related to the fact that the conventional notion of distance breaks down somewhere around the Planck distance, $L_p = \sqrt{\frac{\hbar G}{c^3}}$. This generalized uncertainty principle (GUP) implies also the correction of the commutation relation in the corresponding Heisenberg algebra [3].

The aforementioned change has also been derived not only in connection with quantum geometry [4], but also in association with the way in which quantum measurements alter the local spacetime metric [5]. The emergence of a GUP in several models has suggested that it could be a very general characteristic shared by several theories [6]. This last statement means that the analysis of the consequences of this kind of GUPs could render important conclusions about the physical implications of the attempts to quantize the gravitational field.

In the present work the relation between one of the possible GUPs and the time evolution of a quantum particle immersed in a homogeneous gravitational field will be analyzed. The motion equations for the momentum and position operators will be solved, and it will be shown that the momentum operator does depend in a novel way upon the mass of the involved particle. This kind of physical characteristics could be, in principle, detected. In other words, one way of confronting against the experiment some of the models around quantum gravity is given by the detection of the dependence upon the mass parameter of the expectation value of the momentum operator.

7.2 GUP and the motion equations in a homogeneous gravitational field

Let us now assume that the Heisenberg algebra takes the form [3, 4, 5]

$$[\hat{x}, \hat{p}] = i\hbar\Big[\Pi + \epsilon(\frac{L_p}{\hbar})^2\hat{p}^2\Big]. \tag{7.1}$$

Concerning this last expression it must be mentioned that ϵ denotes a constant, additionally we have considered a one–dimensional system, of course, Π denotes the unit operator. The reason for this last restriction stems from that fact that up to now there is no higher–dimensional generalization of (1) [3].

Consider now a very simple situation, namely a quantum particle of mass m obeys the following Hamiltonian

$$H = \frac{p^2}{2m} - mgx. \tag{7.2}$$

Clearly the involved gravitational potential is related to a homogeneous field.

This is a very simple situation and it may be argued that it provides a very rough approximation to any realistic physical scenario. Nonetheless, we may justify the use of such a simple gravitational field as follows: the emergence of GUPs is a dynamical phenomenon, namely, it is related to the existence of fluctuations of the background metric [4]. Hence, we may assume that in our context the background metric has been approximated by a weak field and slow motion limit, such that we may consider a Newtonian situation, and introduce the effects of the fluctuations of the metric, upon the dynamics of a nonrelativistic particle, taking into account a GUP in the motion equation for the corresponding operators.

Clearly, the assumption of a Newtonian limit is a very stringent restriction. Nonetheless, we may introduce it as the roughest approximation for the background geometry (but the one allows us to deduce an analytical solution to the motion equations), and consider the presence of GUP as quantum gravity corrections, associated with fluctuations of the spacetime geometry, that in our case has an average geometry provided by a homogeneous Newtonian gravitational field. In other words, the fluctuations in the geometry are introduced only in the modifications that they provoke in the Heisenberg algebra.

Of course, a more realistic scenario has to consider a better approximation for the background geometry, but the present approach will give us a taste of the conceptual problems that might emerge if we consider, simultaneously, a gravitational field and a GUP, as fundamental elements in the time evolution of a quantum particle.

Hence, in the Heisenberg picture [7] the motion equations for the momentum and position operators read

$$\frac{d\hat{p}}{dt} = mg\Big[\Pi + \epsilon(\frac{L_p}{\hbar})^2\hat{p}^2\Big], \tag{7.3}$$

$$\frac{d\hat{x}}{dt} = \frac{1}{m}\Big[\hat{p} + \epsilon(\frac{L_p}{\hbar})^2\hat{p}^3\Big]. \tag{7.4}$$

The solutions to these equations are, to first order in $\epsilon\left(\frac{L_p}{\hbar}\right)^2$,

$$\hat{p}(t) = mgt\Pi + \hat{p}(0) + mg\epsilon\Big(\frac{L_p}{\hbar}\Big)^2\Big[\frac{(mg)^2}{3}t^3\Pi + mgt^2\hat{p}(0) + t\hat{p}(0)^2\Big], \tag{7.5}$$

$$\hat{x}(t) = \frac{gt^2}{2}\Pi + \frac{t}{m}\hat{p}(0) + \hat{x}(0) + g\epsilon\Big(\frac{L_p}{\hbar}\Big)^2\Big[\frac{(mg)^2}{3}t^4\Pi + \frac{4}{3}mgt^3\hat{p}(0) + 2t^2\hat{p}(0)^2\Big]. \tag{7.6}$$

Here $\hat{p}(0)$ and $\hat{x}(0)$ are the initial momentum and position operators, respectively, which satisfy the commutation relation

$$[\hat{x}(0), \hat{p}(0)] = i\hbar\Big[\Pi + \epsilon(\frac{L_p}{\hbar})^2\hat{p}(0)^2\Big]. \tag{7.7}$$

Resorting to the aforementioned solutions, and to the last commutation relation, it is readily seen that

$$[\hat{x}(t), \hat{p}(t)] = i\hbar\Big[\Pi + \epsilon(\frac{L_p}{\hbar})^2\hat{p}(t)^2\Big]. \quad (7.8)$$

From the last expression we have that our GUP is always fulfilled. At this point it is noteworthy to comment that one of the consequences of a GUP like (1) is the fact that no physical state is a position eigenstate [3]. Notice that in our approach we have not considered position eigenstates, instead, we deal with position and momentum operators, i.e., we use Heisenberg picture.

Consider now a quantum system whose vector state $|\alpha>$ is an eigenvector of the momentum operator, for our GUP this condition is, mathematically, consistent [3].

$$\frac{<\alpha|\hat{p}(t)|\alpha>}{m} = \frac{P(0)}{m} + gt\Big\{1 + \epsilon\Big(\frac{L_p}{\hbar}\Big)^2\Big[\frac{(mgt)^2}{3}\Pi + mgtP(0) + P(0)^2\Big]\Big\}. \quad (7.9)$$

Here $P(0) =< \alpha|\hat{p}(0)|\alpha >$. At this point it is noteworthy to comment that we have imposed no restriction upon $|\alpha>$, therefore we may consider that it is related to the center of mass of a macroscopic body. Clearly (9) contains a novel mass dependence of the expection value. Indeed, $< \alpha|p(t)|\alpha > /m$ is not mass independent. In the limit $\epsilon \rightarrow 0$, the last two expressions, show that the Newtonian situation is recovered.

Here we must add that there are further generalizations of (1) [3] which include a term with the form $\beta(x/L_p)^2$, and we could wonder about the possible effects of this kind of terms on the dynamics of our particle. Taking a look at (3) we may understand (on classical grounds) the last term on the right–hand side as an additional force, and the introduction of $\beta \neq 0$ will not cancel, in the general case, the effects of $\epsilon(L_p/\hbar)^2p^2$. In other words, a deeper modification of the Heisenberg algebra would have similar consequences upon the dynamics of our quantum system.

7.3 Conclusions

We have considered a quantum particle moving in a region where a homogeneous gravitational field is present, this field is introduced as an average geometry, and the fluctuations of the geometry have been introduced only in context of the modifications that they provoke in the commutation relations. Assuming a GUP, it was proved that $< \alpha|p(t)|\alpha > /m$ is, in the general case, not mass independent, the consequences of this last fact could be far–reaching.

Clearly, expression (9) shows that WEP is not fulfilled. Here, following [8], WEP reads: *if an uncharged test body is placed at an initial event in spacetime and given an initial velocity then its subsequent trajectory will be independent of its internal structure and composition.* In other words, the presence of a

GUP may be understood as an additional force in the dynamics of our particle. This last remark may be rephrased stating that one way of confronting against the experiment some of the models around quantum gravity is given by the detection of the dependence upon the mass parameter of the expectation value of the momentum operator.

Of course, we must also address the issue concerning the feasibility of the detection of this type of effects. From (9) it is readily seen that the detection depends, critically, upon the magnitude of the term $\epsilon\left(\frac{L_p}{\hbar}\right)^2$.

Here we face, from the very outset severe problems:

Firstly, the magnitude of ϵ, which here is a free parameter, can not be predicted from model–independent arguments. In other words, in order to have a value for this parameter we must consider a particular model [3], either string theory [9] (where the aforementioned parameter is related to the Regge slope), quantum geometry [4] (here the parameter is a function of an upper bound associated to the proper acceleration experienced by massive particles along their worldlines), etc., etc.

Secondly, the current experiments in the realm of WEP are almost always performed resorting to classical systems [10], and the present approach requires the test of WEP employing a quantum particle.

Summing up, the connection with the experiment requires a more careful analysis, and the feasibility of the possible experimental proofs needs a deeper study. The possible violation of WEP, as a consequence of the presence of a GUP, could seem a very strong drawback of the model. Nevertheless, quantum gravity effects imply the violation of some very well accepted symmetries , for instance, the possible emergence of a deformed dispersion relation in the study of photon propagation [11] (as a consequence, for example, of the polymer–like structure of spacetime) renders the breakdown of Lorentz symmetry [12]. This last remark allows us to contemplate the violation of WEP, stemming from quantum gravity effects, without any surprise.

Finally, at this point it is noteworthy to comment that the concept of metric theory of gravity (see page 22 of [8]) relies upon the validity of the Weak Equivalence Principle (WEP), and therefore the possible consequences of the present analysis in the context of metric theories could be also interesting.

Acknowledgments

The author would like to thank A. A. Cuevas–Sosa for his help.

Bibliography

[1] D. Amati, M. Ciafaloni, and G. Veneziano, *Phys. Lett.* **B197**, 81 (1987).

[2] D. Amati, M. Ciafaloni, and G. Veneziano, *Int. J. Mod. Phys.* **A3**, 1615 (1988).

[3] A. Kempf, G. Mangano, and R. B. Mann, *Phys. Rev.* **D52**, 1108 (1995).

[4] S. Capozziello, G. Lambiase, and G. Scarpetta, *Int. J. Theor. Phys.* **39**, 15 (2000).

[5] D. V. Ahluwalia, *Phys. Lett.* **B339**, 301 (1994); *Phys. Lett.* **A275**, 31 (2000).

[6] R. J. Adler and D. I. Santiago, *LANL archive gr–qc/9904026.*

[7] J. J. Sakurai, *Modern Quantum Mechanics*, Addison-Wesley Publishing Company, Reading, Mass. (1995).

[8] C. M. Will, *Theory and experiment in gravitational physics*, Cambridge University Press, Cambridge, (1993).

[9] M. Maggiore, *Phys. Rev.* **D49**, 5182 (1994).

[10] C. M. Will, *Living Rev. Rel.* **4**, 4 (2001).

[11] G. Amelino–Camelia, J. Ellis, N. E. Mavromatos, D. V. Nanopoulos, and S. Sarkar, *Nature* **393** (London), 763 (1998); R. Gambini and J. Pullin, *Phys. Rev.* **D59**, 124021 (1999).

[12] D. Sudarsky, L. F. Urrutia, and H. Vucetich, "New Observational Bounds to Quantum Gravity Signals", *LANL, gr–qc/0204027.*

In: Relativity, Gravitation, Cosmology
Editor: V. Dvoeglazov and A. Espinoza Garrido, pp. 97-108
ISBN 1-59033-981-9

Chapter 8

The Seiberg-Witten Map in Noncommutative Field Theory: an Alternative Interpretation

Subir Ghosh
Physics and Applied Mathematics Unit,
Indian Statistical Institute,
203 B. T. Road, Calcutta 700108, India.

Abstract

In this article, an alternative interpretation of the Seiberg-Witten map in noncommutative field theory is provided. We show that the Seiberg-Witten map can be induced in a geometric way, by a field dependent co-ordinate transformation that connects noncommutative and ordinary space-times. Furthermore, in continuation of our earlier works, it has been demonstrated here that the above (field dependent co-ordinate) transformations are present in a gauge fixed version of the relativistic spinning particle model, embedded in the Batalin-Tyutin extended space. We emphasize that the space-time non-commutativity emerges naturally from the particle *spin* degrees of freedom. Contrary to similarly motivated works, the noncommutativity is not imposed here in an *ad-hoc* manner.
Keywords: Seiberg-Witten map, non-commutative space-time, spinning particle.
PACS Numbers: 02.40.Gh; 11.10.Ef; 11.90.+t

8.1 Introduction

Historically the Non-commutative (NC) spacetime was introduced by Snyder [1] as a regularization to tame the short distance singularities, inherent in a Quantum Field Theory (QFT). In some sense, this extra structure might appear as a natural generalization of the phase space non-commutativity in quantum mechanics. The softening of the divergence is obviously because NC in spacetime can introduce a lower bound in the continuity of spacetime, just as $\hbar$ does in the quantum phase space. The advantage of NC as a regularization is that the computational scheme requires very little changes from the ordinary spacetime and in some forms of NC [1], (for more recent works see [2, 3, 4]), manifest Lorentz invariance can be maintained.[1] All the same, the idea of Snyder [1] lost popularity due to the success of renormalization techniques in QFT. Also, now we know [6] that noncommutativity is not of much use as a regulating scheme.

However, in more recent times, NCQFT has matured as an area of intense research activity [6]. It has been established by Seiberg and Witten [7] that the existence of non-commutativity in (open) string boundaries in the presence of a constant two-form Neveu-Schwarz field results in NC D-branes to which the open string endpoints are attached. (Hamiltonian formulation of the above open string boundary noncommutativity is discussed in [8].) This renders QFTs living on the D-brane noncommutative. The authors in [7] also propose an explicit mapping between NC and ordinary space-time dynamical variables. This celebrated map goes by the name of the Seiberg-Witten Map (SWM) [7]. Actually the SWM is the culmination of a much deeper understanding between the connection of QFTs in NC and ordinary space-time. It has been shown in [7] that the appearance of NCQFT is dependent on the choice of regularization and in fact a QFT in ordinary spacetime and an NCQFT both can describe the same underlying QFT. At least to the lowest non-trivial order in $\theta_{\mu\nu}$[2], the non-commutativity parameter,

$$[x_\mu, x_\nu] = i\theta_{\mu\nu}, \tag{8.1}$$

the SWM can be exploited to convert a NCQFT to its counterpart living in ordinary space-time, in which the effects of non-commutativity appear as local interaction terms, supplemented by $\theta_{\mu\nu}$. In the more popular form of NCQFT, $\theta_{\mu\nu}$ is taken to be constant. This can lead to very striking signatures in particle physics phenomenology in the form of Lorentz symmetry breakdown, new interaction vertices etc. [10].

The SWM plays a pivotal role in our understanding of the NCQFT by directly making contact between NCQFT and QFT in ordinary space-time. This has led to new results in the form of axial anomaly [11] in an NC interacting theory of fermions coupled to gauge fields. The previous results [12] are incomplete as they do not conform to the SWM. Furthermore, the SWM is crucial in generalizing [13] to NC space-time, the duality between Maxwell-Chern-Simons and Self-Dual

---

[1] The ideas of [3] is discussed in a field theoretic setting in [5].

[2] The SWM for higher orders in θ are discussed in [9].

theories in 2+1-dimensions [14]. In the context of bosonization of the massive Thirring model in 2+1-dimensions [15], SWM reveals that the resulting theory is different from the NC self-dual model [16], in contrast to ordinary space-time.

However, as it stands, the SWM is linked exclusively to NC *gauge* theory, since the original derivation of the SWM [7] hinges on the concept of identifying gauge orbits in NC and ordinary space-times. In the explicit form of the SWM [7], the non-commutativity of the *space-time* in which the NC gauge field lives, is not manifest at all since the map is a relation between the NC and ordinary gauge fields and gauge transformation parameters, all having ordinary space-time coordinates as their arguments.

On the other hand, possibly it would have been more natural to consider first a map between NC and ordinary space-time and subsequently to induce the SWM through the change in the space-time argument of the gauge field from the ordinary to NC one. Precisely this type of a geometrical reformulation of the SWM is proposed in the present work.

The paper is organized as follows: **Section II** describes the motivation and an outline of our ideas. The SWM is analysed in a different light in **Section III**. This constitutes our alternative interpretation of the SWM. **Section IV** is devoted to a discussion on the Batalin-Tyutin extension and the ordinary and noncommutative spacetime mapping. The Batalin-Tyutin extension and SWM are bridged through the process of gauge fixing in **Section V**. The paper ends with a conclusion and future prospects in **Section VI**.

8.2 Motivation and an outline of the present work

In the canonical quantization prescription, the Poisson Bracket algebra is elevated to quantum commutator algebra by the replacement

$$\{A, B\} \rightarrow \frac{1}{i}[\hat{A}, \hat{B}].$$

But presence of *constraints* may demand a modification in the Poisson Bracket algebra, leading to the Dirac Bracket algebra [17], which are subsequently identified to the commutators,

$$\{A, B\}_{DB} \rightarrow \frac{1}{i}[\hat{A}, \hat{B}].$$

However, complications in this formalism can arise, (in particular in case of non-linear constraints), where the Dirac Bracket algebra itself becomes operator valued. To overcome this, Batalin and Tyutin [18] have developed a systematic scheme in which all the physical variables are mapped in an extended canonical phase space, consisting of auxiliary degrees of freedom besides the physical ones, with all of them enjoying canonical free Poisson Bracket algebra. In this formalism, the ambiguity of using (operator valued) Dirac Brackets as quantum commutators does not arise.

In the spinning particle model [19] the canonical $\{x_\mu, x_\nu\} = 0$ Poisson Bracket changes to an operator valued Dirac Bracket,

$$\{x_\mu, x_\nu\}_{DB} = -\frac{S_{\mu\nu}}{M^2}, \tag{8.2}$$

due to the presence of constraints. In the above, the dynamical variable $S_{\mu\nu}$ represents the spin angular momentum and M is the mass of the particle. This forces us to exploit the Batalin-Tyutin prescription [18].

In a recent paper [3], we have constructed a mapping of the form,

$$\{x_\mu, x_\nu\} = 0 \quad , \quad x_\mu \to \hat{x}_\mu \quad ; \quad \{\hat{x}_\mu, \hat{x}_\nu\} = \theta_{\mu\nu}, \tag{8.3}$$

which bridges the noncommutative and ordinary space-times. Note that $\hat{x}_\mu$ lives in the Batalin-Tyutin [18] extended space and is of the generic form $\hat{x}_\mu = x_\mu + X_\mu$, where X_μ consists of physical and auxiliary degrees of freedom. Explicit expressions for X_μ are to be found later [3].

This space-time map induces in a natural way the following map between noncommutative and ordinary degrees of freedom,

$$\lambda(x) \to \lambda(\hat{x}) \to \hat{\lambda}(x) \quad , \quad A_\mu(x) \to A_\mu(\hat{x}) \to \hat{A}_\mu(x). \tag{8.4}$$

Here $\hat{\lambda}$ and $\hat{A}_\mu$ are the NC counterparts of λ and A_μ, the abelian gauge transformation parameter and the gauge field respectively and $\hat{x}_\mu$ and x_μ are the NC and ordinary space-time co-ordinates.

On the other hand, there also exists the SWM [7] which interpolates between noncommutative and ordinary variables,

$$\lambda(x) \to \hat{\lambda}(x) \quad , \quad A_\mu(x) \to \hat{A}_\mu(x). \tag{8.5}$$

It is only logical that the above two schemes ((8.3-8.4) and 8.5) can be related. In the present work we have precisely done that. The formulation [3] (8.3-8.4) being the more general one, we have explicitly demonstrated how it can be reduced to the SWM [7], in a particular gauge. This incidentally demonstrates the correctness of the procedure. The above idea was hinted in [3].[3]

In this context, let us put the present work in its proper perspective. Recently a number of works have appeared with the motivation of recovering the SWM in a geometric way, without invoking the gauge theory principles [20]. However, the noncommutative feature of the space-time plays no direct role in the above mentioned re-derivations of the SWM, with non-commutativity just being postulated in an *ad hoc* way. In the present work, we have shown how one can construct a noncommutative sector inside an extended phase space, in a relativistically covariant way. More importantly, we have shown explicitly how this generalized map can be reduced to the SWM under certain approximations. Interestingly, this extended space is physically significant and well studied: It is

[3]The present analysis being classical, (non)commutativity is to be interpreted in the sense of Poisson or Dirac Brackets.

the space of the relativistic spinning particle [3, 19]. Hence it might be intuitively appealing to think that the NC space-time is endowed with spin degrees of freedom, as compared to the ordinary configuration space, since the spin variables directly generate the NC [5]. The analogue of the gauge field is also identified inside this phase space, without any need to consider external fields. This situation is to be contrasted with the NC arising from the background magnetic field in the well known Landau problem [6] of a charge moving in a plane in the presence of a strong, perpendicular magnetic field.

8.3 Seiberg-witten map: an alternative interpretation

The genesis of the SWM is the observation [7] that the non-commutativity in string theory depends on the choice of the regularization scheme: it appears in *e.g.* point-spitting regularization whereas it does not show up in Pauli Villars regularization. This feature, among other things, has prompted Seiberg and Witten [7] to suggest the map connecting the NC gauge fields and gauge transformation parameter to the ordinary gauge field and gauge transformation parameter. The explicit form of the SWM [7], for abelian gauge group, to the first non-trivial order in the NC parameter $\theta_{\mu\nu}$ is the following,

$$\hat{\lambda}(x) = \lambda(x) + \frac{1}{2}\theta^{\mu\nu}A_\nu(x)\partial_\mu\lambda(x) \ ,$$

$$\hat{A}_\mu(x) = A_\mu(x) + \frac{1}{2}\theta^{\sigma\nu}A_\nu(x)F_{\sigma\mu}(x) + \frac{1}{2}\theta^{\sigma\nu}A_\nu(x)\partial_\sigma A_\mu(x). \qquad (8.6)$$

The above relation (9.1) is an $O(\theta)$ solution of the general map [7],

$$\hat{A}_\mu(A + \delta_\lambda A) = \hat{A}_\mu(A) + \hat{\delta}_{\hat{\lambda}}\hat{A}_\mu(A), \qquad (8.7)$$

which is based on identifying gauge orbits in NC and ordinary space-time.

First let us show that it is indeed possible to re-derive the SWM using geometric objects. We rewrite the SWM (9.1) in the following way,

$$\hat{\lambda}(x) = \lambda(x) + \frac{1}{2}\{\delta_f[\lambda(x)] - (\lambda(x') - \lambda(x))\} = \lambda(x) + \delta_f[\lambda(x)] \ , \qquad (8.8)$$

$$\hat{A}_\mu(x) = A_\mu(x) + \{\delta_f[A_\mu(x)] - (A_\mu(x') - A_\mu(x))\} = A_\mu(x) + A'_\mu(x) - A_\mu(x'). \qquad (8.9)$$

In the above we have defined,

$$x'_\mu = x_\mu - f_\mu \ , \ A'_\mu(x') = \frac{\partial x^\nu}{\partial x'^\mu}A_\nu(x) \ , \ \lambda'(x') = \lambda(x) \ ,$$

$$f^\mu \equiv \frac{1}{2}\theta^{\mu\nu}A_\nu \ . \qquad (8.10)$$

Here f^μ is the field dependent space-time translation parameter and δ_f constitutes the Lie derivative connected to f^μ,

$$\delta_f[\lambda(x)] = \lambda'(x) - \lambda(x) = -(\lambda(x') - \lambda(x)) = f^i \partial_i \lambda,$$

$$\delta_f[A_\mu(x)] = A'_\mu(x) - A_\mu(x) \ . \qquad (8.11)$$

This shows that the NC gauge parameter ($\hat{\lambda}$) and gauge field ($\hat{A}^\mu$) are derivable from the ordinary one by making a *field dependent* space-time translation f^μ [21]. One can check that the NC gauge transformation of $\hat{A}_\mu(x)$ is correctly reproduced by considering,

$$\hat{\delta}\hat{A}_\mu(x) = \delta(A_\mu(x) + \frac{1}{2}\theta^{\sigma\nu} A_\nu(x) F_{\sigma\mu}(x) + \frac{1}{2}\theta^{\sigma\nu} A_\nu(x) \partial_\sigma A_\mu(x)) \ , \qquad (8.12)$$

where $\delta A_\mu(x) = \partial_\mu \lambda(x)$ is the gauge transformation in ordinary space-time. Hence, if expressed in the form (8.9), the SWM, (at least to $O(\theta)$), can be derived in a geometrical way, without introducing the original identification (8.7) obtained from the viewpoint of a matching between NC and ordinary gauge invariant sectors. Also note that the gauge field $A_\mu(x)$ is treated here just as an ordinary vector field, without invoking any gauge theory properties. This constitutes the first part of our result.

Returning to our starting premises, are we justified in making an identification between $\hat{x}_\mu$ in (8.3)-(8.4) and x'_μ introduced in (8.8)-(8.10), because this relation can connect NC and ordinary space-time. Naively, a relation of the form, $x'_\mu = x_\mu - f_\mu(x)$ can not render the x'-space noncommutative, since the right hand side of the equation apparently comprises of commuting objects only. In our subsequent discussion we will show how this surmise can be made meaningful and will return to this point at the end.

8.4 Batalin-tyutin extension: $x_\mu \rightarrow \tilde{x}_\mu$ mapping

We start by considering a larger space having inherent NC. Such a space, which at the same time is physically appealing, is that of the Nambu-Goto model of relativistic spinning particle [19, 3]. Here the situation is somewhat akin to the open string boundary NC such that the role of Neveu-Schwarz field is played by here by the spin degrees of freedom. The Lagrangian of the model in 2+1-dimensions [19, 3] is,

$$L = [M^2 u^\mu u_\mu + \frac{J^2}{2}\sigma^{\mu\nu}\sigma_{\mu\nu} + MJ\epsilon^{\mu\nu\lambda} u_\mu \sigma_{\nu\lambda}]^{\frac{1}{2}} \ , \qquad (8.13)$$

$$u^\mu = \frac{dx^\mu}{d\tau} \ , \quad \sigma^{\mu\nu} = \Lambda_\rho{}^\mu \frac{d\Lambda^{\rho\nu}}{d\tau} = -\sigma^{\nu\mu} \ , \qquad (8.14)$$
$$\Lambda_\rho{}^\mu \Lambda^{\rho\nu} = \Lambda^\mu{}_\rho \Lambda^{\nu\rho} = g^{\mu\nu} \ , \quad g^{00} = -g^{ii} = 1.$$

Here $(x^\mu \ , \ \Lambda^{\mu\nu})$ is a Poincare group element and also a set of dynamical variables of the theory.

In a Hamiltonian formulation, the conjugate momenta are,

$$P^\mu = \frac{\partial L}{\partial u_\mu} = L^{-1}[M^2 u^\mu + \frac{MJ}{2}\epsilon^{\mu\nu\lambda}\sigma_{\nu\lambda}] \ , \tag{8.15}$$
$$S^{\mu\nu} = \frac{\partial L}{\partial \sigma_{\mu\nu}} = \frac{L^{-1}}{2}[J^2\sigma^{\mu\nu} + MJ\epsilon^{\mu\nu\lambda}u_\lambda].$$

The Poisson algebra of the above phase space degrees of freedom are,

$$\{P^\mu, x^\nu\} = g^{\mu\nu} \ , \ \{P^\mu, P^\nu\} = 0 \ , \ \{x^\mu, x^\nu\} = 0 \ , \ \{\Lambda^{0\mu}, \Lambda^{0\nu}\} = 0 \ , \tag{8.16}$$

$$\{S^{\mu\nu}, S^{\lambda\sigma}\} = S^{\mu\lambda}g^{\nu\sigma} - S^{\mu\sigma}g^{\nu\lambda} + S^{\nu\sigma}g^{\mu\lambda} - S^{\nu\lambda}g^{\mu\sigma} \ , \tag{8.17}$$
$$\Lambda^{0\mu}, S^{\nu\sigma}\} = \Lambda^{0\nu}g^{\mu\sigma} - \Lambda^{0\sigma}g^{\mu\nu} \ .$$

The full set of constraints are,

$$\Psi_1 \equiv P^\mu P_\mu - M^2 \approx 0 \ , \ \Psi_2 \equiv S^{\mu\nu}S_{\mu\nu} - 2J^2 \approx 0, \tag{8.19}$$

$$\Theta_1^\mu \equiv S^{\mu\nu}P_\nu \ , \ \Theta_2^\mu \equiv \Lambda^{0\mu} - \frac{P^\mu}{M} \ , \ \mu = 0, \ 1, \ 2 \ , \tag{8.20}$$

out of which Ψ_1 and Ψ_2 give the mass and spin of the particle respectively.[4] In the Dirac constraint analysis [17], these are termed as First Class Constraints (FCC), having the property that they commute with *all* the constraints on the constraint surface and generate gauge transformations. The set Θ_2^μ is put by hand [3], to restrict the number of angular co-ordinates.

The non-commuting set of constraints $\Theta_\alpha^\mu \ , \alpha = 1, 2$, termed as Second Class Constraints (SCC) [17], modify the Poisson Brackets (8.16) to Dirac Brackets [17], defined below for any two generic variables A and B,

$$\{A, B\}_{DB} = \{A, B\} - \{A, \Theta_\alpha^\mu\}\Delta_{\mu\nu}^{\alpha\beta}\{\Theta_\beta^\nu, B\}, \tag{8.21}$$

$$\{\Theta_\alpha^\mu, \Theta_\beta^\nu\} \equiv \Delta_{\alpha\beta}^{\mu\nu} \ , \ \alpha, \beta = 1, 2 \ , \ \Delta_{\alpha\beta}^{\mu\nu}\Delta_{\nu\lambda}^{\beta\gamma} = \delta_\alpha^\gamma \delta_\lambda^\mu \ . \tag{8.22}$$

$\Delta_{\alpha\beta}^{\mu\nu}$ is non-vanishing even on the constraint surface. The main result, relevant to us, is the following Dirac Bracket [19, 3],

$$\{x_\mu, x_\nu\}_{DB} = -\frac{S_{\mu\nu}}{M^2} \rightarrow \{\hat{x}_\mu, \hat{x}_\nu\} = \theta_{\mu\nu}. \tag{8.23}$$

This is the non-commutativity that occurs naturally in the spinning particle model. Our aim is to express this NC co-ordinate $\hat{x}_\mu$ in the form $\hat{x}_\mu = x_\mu - f_\mu$, with the identification between $\theta_{\mu\nu}$ and $S_{\mu\nu}$. This is indicated in the last equality in (8.23). In the quantum theory, this will lead to the NC space-time (8.3).

[4] Note that instead of Ψ_2 as above, one can equivalently use $\Psi_2 \equiv \epsilon^{\mu\nu\lambda}S_{\mu\nu}P_\lambda - MJ$, which incidentally defines the Pauli Lubanski scalar.

This motivates us to the Batalin-Tyutin quantization [18] of the spinning particle [3]. For a system of irreducible SCCs, in this formalism [18], the phase space is extended by introducing additional BT variables, ϕ_a^α, obeying

$$\{\phi_\mu^\alpha, \phi_\nu^\beta\} = \omega_{\mu\nu}^{\alpha\beta} = -\omega_{\nu\mu}^{\beta\alpha} \quad , \quad \omega_{\mu\nu}^{\alpha\beta} = g_{\mu\nu}\epsilon^{\alpha\beta} \ , \ \epsilon^{12} = 1. \tag{8.24}$$

where the last expression is a simple choice for $\omega_{\mu\nu}^{\alpha\beta}$. The SCCs Θ_α^μ are modified to $\tilde{\Theta}_\alpha^\mu$ such that they become FCC,

$$\{\tilde{\Theta}_\alpha^\mu(q,\phi), \tilde{\Theta}_\beta^\nu(q,\phi)\} = 0 \ ; \ \tilde{\Theta}_\alpha^\mu(q,\phi) = \Theta_\alpha^\mu(q) + \Sigma_{n=1}^\infty \tilde{\Theta}_\alpha^{\mu(n)}(q,\phi) \ ; \ \tilde{\Theta}^{\mu(n)} \approx O(\phi^n), \tag{8.25}$$

with q denoting the original degrees of freedom. Let us introduce the gauge invariant variables $\tilde{f}(q)$ [18] corresponding to each $f(q)$, so that $\{\tilde{f}(q), \tilde{\Theta}_\alpha^\mu\} = 0$

$$\tilde{f}(q,\phi) \equiv f(\tilde{q}) = f(q) + \Sigma_{n=1}^\infty \tilde{f}(q,\phi)^{(n)}, \tag{8.26}$$

which further satisfy [18],

$$\{q_1, q_2\}_{DB} = q_3 \ \rightarrow \{\tilde{q}_1, \tilde{q}_2\} = \tilde{q}_3 \ , \ \tilde{0} = 0. \tag{8.27}$$

It is now clear that our target is to obtain $\tilde{x}_\mu$ for x_μ. Explicit expressions for $\tilde{\Theta}^{\mu(n)}$ and $\tilde{f}^{(n)}$ are derived in [18].

Before we plunge into the BT analysis, the reducibility of the SCCs Θ_1^μ (*i.e.* $P_\mu\Theta_1^\mu{=}0$) [19, 3] is to be removed [22] by introducing a canonical pair of auxiliary variables ϕ and π that satisfy $\{\phi,\pi\} = 1$ and PB commute with the rest of the physical variables. The modified SCCs that appear in the subsequent BT analysis are as shown below:

$$\Theta_1^\mu \equiv S^{\mu\nu}P_\nu + k_1P^\mu\pi \quad ; \quad \Theta_2^\mu \equiv (\Lambda^{0\mu} - \frac{P^\mu}{M}) + k_2(\Lambda^{0\mu} + \frac{P^\mu}{M})\phi \ , \tag{8.28}$$

where k_1 and k_2 denote two arbitrary parameters. Since the computations are exhaustively done in [3] they are not repeated here. The results are the following:

$$\tilde{x}_\mu = x_\mu + [S_{\nu\mu} + 2k_1\pi g_{\nu\mu}](\phi^1)^\nu + \mathcal{R}_{1\mu\nu}(\phi^2)^\nu + higher - \phi - terms \ , \tag{8.29}$$

$$\{\tilde{x}_\mu, \tilde{x}_\nu\} = -\frac{\tilde{S}_{\mu\nu}}{M^2} \ , \ \tilde{S}_{\mu\nu} = S_{\mu\nu} + \mathcal{R}_{2(\alpha)\mu\nu\lambda}\phi^{(\alpha)\lambda} + higher - \phi - terms \ , \tag{8.30}$$

where the expressions for $\mathcal{R}$ are straightforward to obtain [3] but are not needed in the present order of analysis. Only it should be remembered that the $\mathcal{R}_1$-term in (8.29) is responsible for the $(\phi^\alpha)^\mu$-free term $-S_{\mu\nu}/(M^2)$ in the $\{\tilde{x}_\mu, \tilde{x}_\nu\}$ bracket in (8.30). Thus the problem that we had set out to solve has been addressed successfully in (8.29), which expresses the NC $\tilde{x}_\mu$ in terms of ordinary x_μ and other variables [3].

8.5 Batalin-tyutin extension in a gauge → The seiberg-witten map

Now comes the crucial part of identification of the present map with the SWM [7]. This means in particular that we have to connect (8.29) to (8.10), since as we have shown before, (8.10) is capable of generating the SWM [7]. We exploit the freedom of choosing gauges according to our convenience, since in the BT extended space $\tilde{\Theta}^{\mu}_{\alpha}$ are FCCs. For instance, the so called unitary gauge, $\phi^{\mu}_1 = 0, \phi^{\mu}_2 = 0$, trivially converts the system back to its original form before the BT extension. Let us choose the following non-trivial gauge,

$$\phi^{\mu}_1 = \frac{M^2}{2}A^{\mu}(x) \quad , \quad \phi^{\mu}_2 = 0, \tag{8.31}$$

where $A^{\mu}(x)$ is some function of x_{μ}, to be identified with the gauge field. Let us also work with terms linear in $A^{\mu}(x)$. Identifying $\tilde{S}_{\mu\nu}/(M)^2 = \theta_{\mu\nu}$ we end up with the cherished mapping,

$$\tilde{x}_{\mu} = x_{\mu} - \frac{1}{2}\theta_{\mu\nu}A^{\nu}(x) + higher - A(x) - terms \ , \tag{8.32}$$

$$\{\tilde{x}_{\mu}, \tilde{x}_{\nu}\}_{DB} = \theta_{\mu\nu} + higher - A(x) - terms \ . \tag{8.33}$$

Note that in the above relations (8.32,8.33), we have dropped the terms containing k_1, an arbitrary parameter [3], considering it to be very small. Also in (8.33) Dirac Bracket reappears since the system is gauge fixed and hence has SCCs. This constitutes the second part of our result.

Finally, two points are to be noted. Firstly, the non-commutativity present here does *not* break Lorentz invariance since there appear no constant parameter with non-trivial Lorentz index to start with. The violation will appear only in the identification of $\tilde{S}_{\mu\nu}$ with (constant) $\theta_{\mu\nu}$. Secondly, (8.29) truly expresses the NC space-time $\tilde{x}_{\mu}$ in terms of ordinary space-time x_{μ}. But x_{μ} becomes NC owing to the Dirac brackets induced by the particular gauge that we fixed in order to reduce our results to the SWM. Obviously, in general, there is no need to fix this particular gauge. This refers to the comment below (8.12).

8.6 Conclusions and future prospects

In conclusion, let us summarize our work. We have shown that the (abelian $O(\theta)$) Seiberg-Witten map can be viewed as a co-ordinate transformation, albeit with field dependent parameters. The duality concept between gauge orbits in noncommutative and ordinary space-times, which was crucial in the original derivation [7], is not applied here. It has been explicitly demonstrated that a noncommutative space-time sector can be constructed in the Batalin-Tyutin extension of the relativistic spinning particle model [3]. Finally, the above mentioned transformation and subsequently a direct connection with the

Seiberg-Witten map is also generated in this model. The present work reveals that noncommutative space-time is endowed with spin degrees of freedom, as compared to the ordinary space-time [5].

As a future work, we plan to develop the quantum theory of the noncommutative spacetime model proposed in [5] whose particle content is analogous to the spinning particles considered here. This theory has a classical conformal invariance, the fate of which will be studied upon quantization. Also, it would be interesting to investigate how the constraints induce the noncommutativity in spacetime coordinates in an operator product expansion approach.

Acknowledgement

It is a pleasure to thank Professor R. Jackiw for helpful correspondence.

Bibliography

[1] H.S. Snyder, *Phys. Rev.* 71 38(1947).

[2] S. Doplicher, K. Fredengagen and J.E. Roberts, *Phys. Lett.* B331 39(1994); C.E.Carlson, C.D. Carrone and N. Zobin, *Non-commutative gauge theory without Lorentz violation, Phys. Rev.* D66 (2002) 075001 (HEP-TH/0206035).

[3] S. Ghosh, *Phys. Rev.* D66 045031(2002).

[4] H. Kase, K. Morita, Y. Okumura and E. Umezawa, *Lorentz-Invariant Non-Commutative Space-Time Based On DFR Algebra*, HEP-TH/0212127.

[5] S. Ghosh, *Modelling a noncommutative two-brane*, hep-th/0212123.

[6] For a review see M.R. Douglas and N.A. Nekrasov, *Rev. Mod. Phys.* 73 977(2002) (arXiv: HEP-TH/ 0106048); R.J. Szabo, HEP-TH/0109162; A. Connes, *Noncommutative Geometry*, Academic Press, 1994.

[7] N. Seiberg and E. Witten, *JHEP* 09(1999)032.

[8] R. Banerjee, B. Chakraborty and S. Ghosh, *Phys. Lett.* B537 340(2002).

[9] Y. Okawa and H. Ooguri, *Phys. Rev.* D64 046009(2001); S. Fidanza, JHEP 0206 016(2002).

[10] X. Calmet et. al., *HEP-PH/0111115*; H. Bozkaya, P. Fischer, H. Grosse, M. Pitschmann, V. Putz, M. Schweda, R. Wulkenhaar, *Space/time noncommutative field theories and causality*, HEP-TH/0209253; M. Chaichian, K. Nishijima and A. Tureanu, *Spin-Statistics and CPT Theorems in Noncommutative Field Theory*, HEP-TH/0209008; T. Tamaki, T. Harada, U. Miyamoto and T. Torii, *Particle velocity in noncommutative space-time, Phys. Rev.* D66 (2002) 105003 (gr-qc/0208002); J.L. Cortes, J. Gamboa, F. Mendez, *Noncommutativity in Field Space and Lorentz Invariance Violation*, J.M. Carmona, *HEP-TH/0207158.*

[11] R. Banerjee and S. Ghosh, *Phys. Lett.* B533 162(2002).

[12] J.M. Gracia-Bondia and C.P. Martin, *Phys. Lett.* B479 321(2000); L.Bonora, M.Schnabl and A.Tomasiello, *Phys. Lett.* B485 311(2000); F. Ardalan and N. Sadooghi, *Int. J. Mod. Phys.* A16 3157(2001).

[13] S. Ghosh, *Phys. Lett.* B 558 245 (2003) (hep-th/0210107); M.B. Cantcheff and P. Minces, *Phys. Lett.* B557 283(2003) (hep-th/0212031); O.F. Dayi, *Phys. Lett.* B560 239(2003).

[14] S. Deser, R. Jackiw and S. Templeton, *Phys. Rev. Lett.* 48(1982)975; *Ann. Phys.* 140 (1982)372; S. Deser and R. Jackiw, *Phys. Lett.* 139B(1984)371.

[15] E.Fradkin and F.A.Schaposnik, *Phys.Lett.* 338B(1994)253; R.Banerjee, *Phys.Lett.* 358B (1995)297.

[16] S. Ghosh, *Phys. Lett.* B563 112(2003).

[17] P.A.M. Dirac, *Lectures on Quantum Mechanics* (Yeshiva University Press, New York, 1964).

[18] I.A. Batalin and I.V. Tyutin, *Int. J. Mod. Phys.* A6 3255(1991).

[19] S. Ghosh, *Phys.Lett.* B338 235(1994); *J. Math. Phys.* 42 5202(2001). For the original works see, A.J. Hanson and T. Regge, *Ann. Phys.* (N.Y.) 87 498(1974); see also A.J. Hanson, T. Regge and C. Teitelboim, *Constrained Hamiltonian System*, Roma, Accademia Nazionale Dei Lincei, (1976).

[20] L. Cornalba, *D-brane physics and noncommutative Yang-Mills theory*, HEP-TH/9909081; A.A. Bichl et. al., *Noncommutative Lorentz symmetry and the origin of the Seiberg-Witten map*, HEP-TH/0108045; B. Jurco and P. Schupp, *Euro. Phys. J.* C14 367(2000); B. Jurco and P. Schupp and J. Wess, *Nucl. Phys.* B584 784(2001); H.Liu, *Nucl. Phys.* B614 305(2001); R. Jackiw, S.-Y. Pi and A.P. Polychronakos, *Noncommutating gauge fields as a Lagrange fluid*, *Phys. Rev. Lett.* 88 111603(2002) (HEP-TH/0206014).

[21] Field dependent transformations have appeared in *eg* R. Jackiw, *Phys. Rev. Lett.* 41, 1635 (1978); R. Jackiw and S.-Y. Pi, HEP-TH/0111122.

[22] R. Banerjee and J. Barcelos-Neto, *Ann. Phys.* 265 134(1998).

In: Relativity, Gravitation, Cosmology
Editor: V. Dvoeglazov and A. Espinoza Garrido, pp. 109-122
ISBN 1-59033-981-9

Chapter 9

Relativistic Brownian Motion

O. Oron and L.P. Horwitz
School of Physics and Astronomy
Raymond and Beverly Sackler Faculty of Exact Sciences
Tel Aviv University, Ramat Aviv 69978, Israel
Also at Department of Physics, Bar Ilan University, Ramat Gan 529000, Israel

Abstract

We solve the problem of formulating Brownian motion in a relativistically covariant framework in 1+1 dimensions. We obtain covariant Fokker-Planck equations with (for the isotropic case) a differential operator of invariant d'Alembert form. Treating the spacelike and timelike fluctuations separately, we show that it is essential to take into account the analytic continuation of "unphysical" fluctuations in order to achieve these results.

9.1 Introduction

Nelson [1], in 1966, constructed the Schrödinger equation from an analysis of Brownian motion by identifying the forward and backward average velocities of a Brownian particle with the real and imaginary parts of a wave function. He pointed out that the basic process involved is defined non-relativistically, and can be used if relativistic effects can be "safely" neglected. The development of a relativistically covariant formulation of Brownian motion could therefore provide some insight into the structure of a relativistic quantum theory.

In recent years, Brownian motion has been applied as a mechanism for the description of irreversible evolution of quantum mechanical states, for example, collapse of wave functions as result of measurement. One finds that the wave function with a stochastic term added to the Schrödinger equation (represented in the Hilbert space of states [2] or in their projective representation [3]) may evolve to a mixed state of alternative outcomes of a measurement with the same probability distribution as given by the calculation of *a priori* (Born) probabilities in the framework of the standard quantum theory[2,3]. One of the motivations for the work of Adler and Horwitz [3] was based on the existence of a statistical mechanics, developed by Adler [4] and Adler and Millard [5] for the description of the equilibrium state of a general class of quantum fields; in this theory, the thermal average of commutator expressions take on the values of those of the familiar complex canonical quantum theory and quantum field theory. The Brownian motion corrections to the standard theories may be thought of as arising from the fluctuations around the equilibrium state. The development of a relativistically covariant theory of Brownian motion could make an extension of these ideas to relativistic quantum theory and quantum field theory more accessible.

The program of stochastic quantization of Parisi and Wu [6] assumes the existence of relativistic wave equations, and applies a statistical approach closely related to path integrals. The Wiener distribution of non-relativistic Brownian motion, as is well known, is associated with (imaginary time) path integral formulations, and one would expect that covariant Brownian motion would similarly be associated with Parisi-Wu stochastic quantization.

Nelson has pointed out that the formulation of his stochastic mechanics in the context of general relativity is an important open question. The Riemannian metric spaces one can achieve, in principle, which arise due to nontrivial correlations between fluctuations in spatial directions, could, in the framework of a covariant theory of Brownian motion, lead to spacetime pseudo-Riemannian metrics in the structure of diffusion and Schrödinger equations. In recent years Morato and Viola [7] have constructed a relativistic quantum equation for a free scalar field. They assumed the existence of a 3D (spatial) diffusion in the comoving frame; it is a non-inertial frame in which the average velocity field of the Brownian particle (current velocity) is zero. In this frame the location of the Brownian particle in space experiences Brownian fluctuations parametrized by the proper time of the comoving observer. They interpreted the solutions of the negative 0-component current velocity with what they called 'rare events', which are time reversed Brownian processes (a peculiarity arising in the relativistic treatment). The equation they achieved this way is approximately the Klein-Gordon equation. It is important to note that in the inertial frame they do not obtain a normal diffusion. This is due to the fact that their process is stochastic only in three degrees of freedom and therefore is not covariant.

In this paper we introduce a new approach to the formulation of relativistic Brownian motion. The process is a straightforward generalization of the standard diffusion to 4 dimensions (where the actual random process is thought of as a 'diffusion' in all 4 directions), existing in an inertial frame. The equation achieved is an exact Klein-Gordon equation. It is a relativistic generalization of Nelson's and will be used in a consequent paper to construct diffusion and

Schrödinger equations with pseudo-Riemannian metrics which follow from the existence of nontrivial correlations between the coordinate random variables.

9.2 The Difficulties in Formulating Relativistic Brownian Motion

Brownian motion, thought of as a series of "jumps" of a particle along its path, necessarily involves an ordered sequence. In the nonrelativistic theory, this ordering is naturally provided by the Newtonian time parameter. In a relativistic framework, the Einstein time t does not provide a suitable parameter. If we contemplate jumps in spacetime, to accommodate a covariant formulation, a possible spacelike interval between two jumps may appear in two orderings in different Lorentz frames. The introduction of proper time as a parameter for the RBP (Relativistic Brownian Process) is not adequate since in this case the second order correlations in the simplest case

$$E(\Delta x_\mu \Delta x^\mu) \equiv \Delta s^2 \propto \Delta s, \tag{9.1}$$

where the first equality is by the definition of proper time and the second equality is due to the Brownian property expressed in Eq.(9.1), and there is an obvious contradiction. We therefore adopt the invariant parameter τ introduced by Stueckelberg [8] in his construction of a relativistically covariant classical and quantum dynamics. For the many body theory, Piron and Horwitz [9] postulated that this parameter is universal, as the Newtonian time in the nonrelativistic theory, and drives the classical particle trajectories $x_i{}^\mu(\tau)$ (worldlines labelled $i = 1, 2, 3, ...N$) through equations of motion, and the evolution of the wave function in the Hilbert space $L^2(R^{4N})$, $\psi_\tau(\{x_i{}^\mu\})$ through the Stueckelberg-Schrödinger equation (the differential form of the action of a one-parameter unitary group with parameter τ).

A second fundamental difficulty in formulating a covariant theory of Brownian motion lies in the form of the correlation function of the random variables of spacetime. The straight forward generalization of the simplest case of the standard non-relativistic Wiener process (where $< dw^i dw^j >= \delta^{ij} dt$) to special relativity, i.e.,

$$< dw_\mu(\tau) dw_\nu(\tau') >= \begin{cases} 0 & \tau \neq \tau' \\ 2\alpha\eta_{\mu\nu} d\tau & \tau = \tau', \end{cases} \tag{9.2}$$

contains the serious problem that $< dw_0(\tau) dw_0(\tau) > < 0$, which is impossible. Brownian motion in spacetime, however, should be a generalization of the non-relativistic problem, constructed by observing the non-relativistic process from a moving frame according to the transformation laws of special relativity. Hence, as a first step, the process taking place in space in the nonrelativistic theory should be replaced by a spacetime process in which the Brownian jumps are spacelike. The pure time (negative) self-correlation does therefore not occur. In order to meet this requirement, we shall use a coordinatization in terms of generalized polar coordinates which assure that all jumps are spacelike. Consider for example a relativistic Brownian probability density of the form $e^{-\frac{\mu^2}{a d\tau}}$, where μ is the invariant spacelike interval of the jump. It is a straightforward generalization of the standard Brownian process in 3D, which is generated by a probability density of the form $e^{\frac{-r^2}{a dt}}$, where r is the rotation invariant (i.e. the vector length) and a is proportional to the diffusion constant. We shall refer to this function as the relativistic Gaussian. As we shall see, a Brownian motion based on purely spacelike jumps does not yield the correct form for an invariant diffusion process. We must therefore consider the possibility as well that, in the framework of relativistic dynamics, there are timelike jumps. The corresponding distribution would be expected to be of the form $e^{-\frac{\sigma^2}{b d\tau}}$, where σ is the timelike interval of these jumps, and b is some constant. By suitably weighting the occurrence of the spacelike process (which we take for our main discussion to be "physical", since its nonrelativistic limit coincides with the usual Brownian motion) and an analytic continuation of the timelike process, We show that one indeed obtains a Lorentz invariant Fokker-Planck equation in which the d'Alembert operator appears in place of the Laplace operator of the 3D Fokker-Planck equation. One may, alternatively, consider the timelike process as "physical" and analytically continue the spacelike ("unphysical") process to achieve a d'Alembert operator with opposite sign.

9.3 Brownian motion in 1+1 dimensions

We consider a Brownian path in $1+1$ dimensions generated by the stochastic differential, of the form:

$$dx^\mu(\tau) = \beta^\mu(x(\tau))d\tau + dw^\mu(\tau), \tag{9.3}$$

where dw is a random process, which is a relativistic generalization of the Wiener process, whose properties will be defined later and β^μ is a deterministic field (the forward drift).

We start by considering the second order term in the series expansion of a function of position of the particle on the world line, $f(x^\mu(\tau) + \Delta x^\mu)$, involving the operator

$$\mathcal{O} = \Delta x^\mu \Delta x^\nu \frac{\partial}{\partial x^\mu}\frac{\partial}{\partial x^\nu}. \tag{9.4}$$

We have remarked that one of the difficulties in describing Brownian motion in spacetime is the possible occurrence of a negative value for the second moment of some component of the Lorentz four vector random variable. If the Brownian jump is timelike, or spacelike, however, the components of the four vector are not independent, but must satisfy the timelike or spacelike constraint. Such constraints can be realized by using parameterizations for the jumps in which they are restricted geometrically to be timelike or spacelike. We now separate the random jumps into space-like jumps and time-like jumps accordingly, i.e., for the spacelike jumps,

$$\Delta w^1 = \pm\mu \cosh\alpha \ , \ \Delta w^0 = \mu \sinh\alpha \tag{9.5}$$

and for the timelike jumps,

$$\Delta w^1 = \sigma \sinh\alpha \ , \ \Delta w^0 = \pm\sigma \cosh\alpha(2.3') \tag{9.6}$$

Here we assume that the two sectors have the same distribution on the hyperbolic variable. We furthermore assume that μ,σ are generated by a relativistic Gaussian distribution, working in a Lorentz frame where the α distribution is assumed at this point, to be independent of the μ, σ and is uniformly distributed on the restricted interval $[-L, L]$ where L is arbitrary large. Therefore in this frame $< \Delta w^\mu >$ is 0 and we pick a normalization such that $< \Delta w^n > \propto \Delta\tau^{\frac{n}{2}}$ so to first order in $\Delta\tau$ the contribution to $< \mathcal{O} >$ comes only from $< \Delta w^\mu \Delta w^\nu >$.

Let us look for the effects of a particle experiencing space-like jumps only. In that case the operator $\mathcal{O}$ takes the following form:

$$\mathcal{O}_{spacelike} = \mu^2[\cosh^2\alpha\frac{\partial^2}{\partial x^2} + 2\sinh\alpha\cosh\alpha\frac{\partial^2}{\partial x\partial t} + \sinh^2\alpha\frac{\partial^2}{\partial t^2}] \tag{9.7}$$

If the particle undergoes time-like jumps only the operator $\mathcal{O}$ takes the following form:

$$\mathcal{O}_{timelike} = \sigma^2[\sinh^2\alpha\frac{\partial^2}{\partial x^2} + 2\sinh\alpha\cosh\alpha\frac{\partial^2}{\partial x\partial t} + \cosh^2\alpha\frac{\partial^2}{\partial t^2}]. \tag{9.8}$$

Since μ, σ and α are random processes, the average value of the operator $\mathcal{O}$ is the sum of the two averages of eq.(2.4) and eq.(2.5). Therefore, a difference between these two averages, leading to the d'Alembertian operator can only be obtained by considering the analytic continuation of the timelike process to the spacelike domain, choosing $\mu^2 = -\sigma^2$. This procedure is analogous to the effect, well-known in relativistic quantum scattering theory, of a physical process in the crossed (t)channel on the observed process in the direct (s) channel. For example, in the LSZ formulation of relativistic scattering in quantum field theory [10], a creation operator in the "in" state may be moved to left of the vacuum expectation value expression for the S-matrix, and an annihilation operator for the "out" state may be moved to the right. The resulting amplitude, identical to the original one in value, represents a process that is unphysical; its total "energy" (the sum of four-momenta squared) now has the wrong sign. Assuming that the S-matrix is an analytic function, one may then analytically continue the energy-momentum variables to obtain the correct sign for the physical process in the new channel. Although we are dealing with an apparently classical process, as Nelson [1] has shown, the Brownian motion problem gives rise to a Schrödinger equation, and therefore contains properties of the differential equations of the quantum theory. We thus see the remarkable fact that one must take into account the physical effect of the analytic continuation of processes occurring in a non-physical, in this case timelike, domain, on the total observed behavior of the system.

In the timelike case, the velocity of the particle $\Delta w^1/\Delta w^0 \leq 1$. We shall here use the dynamical association of coordinate increments with energy and momentum

$$E = M\frac{\Delta w^0}{\Delta\tau} \quad p = M\frac{\Delta w^1}{\Delta\tau}, \tag{9.9}$$

so that

$$\sigma^2 = \left(\frac{\Delta\tau}{M}\right)^2(E^2 - p^2), \tag{9.10}$$

where M is a parameter of dimension mass associated with the Brownian particle. It then follows that $E^2 - p^2 = \left(\frac{M}{\Delta\tau}\right)^2\sigma^2 > 0$. For the spacelike case, where $p/E > 1$, we may consider the transformation to an imaginary representation $E \to iE'$ and $p \to ip'$, for E', p' real[1], so the relation $p'/E' > 1$ remains, but $E^2 - p^2 = p'^2 - E'^2 > 0$. In this case, we take the analytic continuation such that the magnitude of σ^2 remains unchanged, but changes sign to become $-\mu^2$ assuring the formation of the d'Alembert operator when the timelike and spacelike fluctuations are added with equal weight (this equality is consistent with the natural assumption, in this case, of an equal distribution between spacelike and timelike contributions). The preservation of the magnitude of the interval reflects the conservation of a mass-like property which remains, as an intrinsic property of the particle, for both spacelike and timelike jumps. As mentioned before one recalls the role of analytic continuation in quantum field theory; for the well known Wick rotation [11], however,only the 0-component is analytically continued and no clear direct physical idea or quantity is associated with it. In the RBP the identification of the imaginary 4-momentum is dynamical in origin. It is due to the Lorentz structure of spacetime, which distinguishes the transitions $\frac{\Delta \mathbf{x}}{\Delta t} > 1$ from those with $\frac{\Delta \mathbf{x}}{\Delta t} < 1$. Though one may object the association of Δx^μ with dynamical momentum (since the instantaneous derivative $\frac{dx^\mu}{d\tau}$ is not defined for a Brownian process) it could be understood as an approximation to a microscopic process, just as it appears in Einstein [12] famous work in 1905, where it is assumed that the Brownian motion is produced by collisions. The conservation of $E^2 - \mathbf{p}^2$ as a real quantity in both timelike and spacelike processes suggests, that it is a physical property which preserves its meaning in both sectors (this suggests further development of the dynamical consequences of this notion).

With these assumptions, the cross-term in hyperbolic functions cancels in the sum, which now takes the form

$$\mathcal{O} = \mu^2\left[\frac{\partial^2}{\partial x^2} - \frac{\partial^2}{\partial t^2}\right] \tag{9.11}$$

[1] This transformation is similar to the continuation $p \to ip'$ in nonrelativistic tunnelling, for which the analytic continuation appears as an instanton.

Taking into account the drift term in (2.1), one then finds the relativistic Fokker-Planck equation

$$\frac{\partial\rho(x,\tau)}{\partial\tau} = \{-\frac{\partial}{\partial x^\mu}\beta^\mu + \langle\mu^2\rangle\frac{\partial}{\partial x^\mu}\frac{\partial}{\partial x_\mu}\}\rho(x,\tau), \tag{9.12}$$

where $\partial/\partial x^\mu$ operates on both β^μ and ρ.

We see that the procedure we have followed, identifying $\sigma^2 = -\mu^2$ and assuming equal weight, permits us to construct the Lorentz invariant d'Alembertian operator, as required for obtaining a relativistically covariant diffusion equation.

To see this symmetry more explicitly, let us define the invariant $\kappa^2 \equiv E_t^2 - p_t^2 \geq 0$ for the timelike case; our requirement is then that $E_s^2 - p_s^2 = -\kappa^2$ for the spacelike case. In the framework of a larger group that includes κ as part of a three vector (E, κ, p), the relation for the timelike case can be considered as the invariant of the group $O(1,2)$, i.e., $E^2 - \kappa^2 - p^2$. The change in sign for the spacelike case yields the invariant $E^2 + \kappa^2 - p^2$, designating the corresponding symmetry (keeping the order of E and p) as $O(2,1)$. These two groups may be thought of as subgroups of $O(2,2)$, where there exists a transformation which changes the sign of metric of the subgroups holding the quantity κ^2 constant. The kinematic constraints we have imposed correspond to setting these invariants to zero (the zero interval in the $2+1$ and $1+2$ spaces).

The constraint we have placed on the relation of the timelike and spacelike invariants derives from the properties of the distribution function and the requirement of obtaining the d'Alembert operator, i.e, Lorentz covariance of the diffusion equation. It appears that in order for the Brownian motion to result in a covariant diffusion equation, the distribution function has a higher symmetry reflecting the necessary constraints. The transformations $E \to iE'$ and $p \to ip'$ used above would then correspond to analytic continuations from one (subgroup) sector to another.

9.4 The Invariance Properties of the Process

Before we continue and formulate the 3+1 dimensional Brownian process, let us investigate the Lorentz invariance of the process and the correlation functions. The averaging operations are actually summations with weights (probability) assigned to each quantity in the sum. Those sums in the continuum are of course, expressed by integrals. If we wish to assign a relativistic Gaussian distribution function then the hypergeometric angle integration is infinite unless we introduce a cutoff. The question then arises whether our process is invariant or not.

We will show that we can use an arbitrary non invariant probability distribution (for example, a cutoff on the hypergeometric angle) and still obtain Lorentz invariant averages, using the imaginary representations of the 'unphysical jumps'. For example, $< \Delta w^\mu >$ stands for a summation with a scalar weight (given by the density) over all the vectors Δw^μ, in the domain. It is therefore a vector. Moreover under the imaginary representation of spacelike increments relative to the timelike ones, Δw^μ is a simple vector function over all spacetime which has the following form:

$$\begin{aligned} \Delta w^\mu &= \Delta w'^\mu \quad , \quad \Delta w'^\mu \text{ timelike} \\ \Delta w^\mu &= i\Delta w'^\mu \quad , \quad \Delta w'^\mu \text{ spacelike,} \end{aligned} \tag{9.13}$$

where the $\Delta w'^\mu$ are all real. The quantity $< \Delta w^\mu >$ (formally written as a discrete sum) is given therefore by:

$$\begin{aligned} < \Delta w^\mu > &= \sum P(\Delta w)\Delta w^\mu = \sum_{\text{timelike}} P'(\Delta w')\Delta w'^\mu + i \sum_{\text{spacelike}} P'(\Delta w')\Delta w'^\mu \\ &= < \Delta w'^\mu >_{\text{timelike}} + i < \Delta w'^\mu >_{\text{spacelike}} \end{aligned} \tag{9.14}$$

where $P(\Delta w)$ (or $P'(\Delta w')$) is the probability(weight) of having the vector Δw (or $\Delta w'$). The two vectors in the last equality in eq.(3.2) are just normal Lorentz vectors. If we now pick a distribution in a given frame for which the average of each of them (independent of the other) is zero then $< \Delta w^\mu >= 0$ is true in all frames since the 0-vector is Lorentz invariant.

Building the second correlation, again with our assumptions of no correlation between spacelike jumps and timelike jumps, we find:

$$\begin{aligned} < \Delta w^\mu \Delta w^\nu > = \sum P(\Delta w)\Delta w^\mu \Delta w^\nu = \sum_{\text{timelike}} P'(\Delta w')\Delta w'^\mu \Delta w'^\nu \\ +i^2 \sum_{\text{spacelike}} P'(\Delta w')\Delta w'^\mu \Delta w'^\nu = \sigma^{\mu\nu}_{\text{timelike}} - \sigma^{\mu\nu}_{\text{spacelike}} \propto \eta^{\mu\nu} D\Delta\tau \end{aligned} \tag{9.15}$$

where $\sigma^{\mu\nu}$ is the correlation tensor in each case. From the definition of dw'^μ (a four vector) it follows that $\sigma^{\mu\nu}$ are real Lorentz tensors. The last equality in eq.(3.3) is a demand that could be achieved for the general 1+n case, by assuming that in that given frame there is an invariant Gaussian distribution where the distribution is uniform in all angles and that there is a cutoff in the hyperbolic angle. The summing of the two non-invariant tensors (each is a result of summation of a different sector) is a Lorentz invariant tensor and therefore it is Lorentz invariant. The higher correlations do not interest us since they are of higher order in ρ and therefore in $\Delta\tau$ and do not contribute to the Fokker-Plank equation.

9.5 The Markov Relation and the Euclidian Gaussian in the 1+1 Case

In developing the previous ideas leading to the formulation of a RBP, we assumed that the probability distribution is consistent with the Markov property phrased in the Chapman-Kolmagorov equation [11]

$$p(x, \tau, y, s) = \int_{\mathcal{R}} p(x, \tau, x', \tau')p(x', \tau', y, s)d^4x' \tag{9.16}$$

the probability of going from x at τ to y at s is the sum over all trajectories starting in x at τ ending in y at s and passing in the intermediate time τ' at all possible points. However, for the relativistic Gaussian it is not clear whether eq.(4.1) holds. Therefore we now consider an alternative process, using the ideas developed previously, resulting eventually in the Klein-Gordon equation. Let us consider a 2D Gaussian process generated by a distribution of the form :

$$p(w, d\tau) = \frac{1}{2\pi D d\tau} exp(\frac{-\Delta {w_0}^2 - \Delta {w_1}^2}{2Dd\tau}) \tag{9.17}$$

It is a Markov process, a standard normalized Wiener process, where D is the diffusion constant. We now use the coordinate representation given in eq.(2.3) and eq.(2.3') for the timelike and spacelike sectors to transform the distribution function in eq.(4.2) in both sectors to :

$$\frac{1}{2\pi D d\tau} exp(\frac{-\mu^2 \cosh 2\alpha}{2Dd\tau}) \tag{9.18}$$

Then, following section 3, using eq.(3.1), we get for the combination of the timelike and spacelike contributions (with the appropriate sign) of the averages of, say, Δw_0,

$$\begin{aligned} < \Delta {w_0}^2 > &= \frac{1}{2\pi D d\tau} \int_0^\infty \int_{-\infty}^\infty \mu^3 exp(\frac{-\mu^2 \cosh 2\alpha}{2Dd\tau}) d\mu d\alpha = \\ &= \frac{1}{\pi} D d\tau \end{aligned} \tag{9.19}$$

where we integrated over μ first, using

$$\int_0^\infty \mu^n exp(-a\mu^2) = \frac{\Gamma(\frac{n+1}{2})}{2a^{(n+1)/2}} \tag{9.20}$$

and then integrated over α using

$$I_2 \equiv \int_{-\infty}^\infty \frac{d\alpha}{\cosh^2 2\alpha} = 1 \tag{9.21}$$

In a similar way one finds that

$$< \Delta {w_1}^2 >= -\frac{1}{\pi} D d\tau \tag{9.22}$$

Since the probability distribution eq.(4.2) is symmetric in Δw_i in each sector $< \Delta w_0 \Delta w_1 >= 0$ as well as the first moments. Therefore we get in this particular frame a d'Alembertian. However, following the results of section 3 it is an invariant result in all Lorentz frames (though in other frames the distribution may not appear to be Gaussian).

9.6 Conclusions and Discussion

We have constructed a relativistic generalization of Brownian motion, using an invariant world-time to order the Brownian fluctuations, and separated consideration of spacelike and timelike jumps to avoid the problems of negative second moments which might otherwise follow from the Minkowski signature.

At equilibrium, where $\partial D/\partial\tau = 0$, the resulting diffusion equation turns into a classical wave equation which, in the absence of a drift term K^μ, is the wave equation for a massless field. An exponentially decreasing distribution in τ of the form $\exp -\kappa\tau$ would correspond to a Klein-Gordon equation for a particle in a tachyonic state (mass squared $-\kappa$). We have considered the spacelike jumps as "physical" since they result in the usual Brownian motion in the nonrelativistic limit. If the timelike jumps were considered as "physical", one would analytically continue the "unphysical" spacelike process. The resulting diffusion equation would have the opposite sign for the d'Alembert operator, and an exponentially decreasing distribution would then result in a Klein-Gordon equation in a timelike particle state.

Nelson [1] has shown that non-relativistic Brownian motion can be associated with a Schrödinger equation. Equipped with the procedures we presented here, constructing relativistic Brownian motion, Nelson's methods can be generalized. One then can construct relativistic equations of Schrödinger (Schrödinger-Stueckelberg) type. The eigenvalue equations for these relativistic forms are also Klein-Gordon type equations. Moreover one can also generalize the case where the fluctuations are not correlated in different directions into the case where correlations exist, as discussed by Nelson [1] for three dimensional Riemannian spaces. In this case the resulting equation is a quantum equation in a curved Riemannian spacetime; as pointed out in ref. [13], the eikonal approximation to the solutions of such an equation contains the geodesic motion of classical general relativity. The medium supporting the Brownian motion may be identified with an "ether" [1] (Nelson [1] has remarked that the self-interaction of charged particles might play the role of the Brownian medium) for which the problem of local Lorentz symmetry is solved. This generalization of Nelson's method will be discussed in a succeeding paper.

In Section 3 we introduced the w manifold which is complex and it is a function of the motion on the real manifold w'. Our macroscopic (physical) equations are written on the real plane of the w manifold. One can then visualize the flow of an event in spacetime similar to a motion of a particle in a bubble chamber. There as the particle moves the bubble condenses, therefore the particle leaves a track. The track itself is not the particle but a result of the actual motion of the particle and its interaction with the gas in the bubble chamber. The extension of the formalism into 3+1 dimension is not straightforward; it is achieved by assigning different weights to the spacelike motion relative to the timelike motion, and will be discussed in a succeeding paper.

Acknowledgements

We would like to thank Linda Reichl, W.C. Schieve and Sungyan Kim at the Ilya Prigogine Center for Statistical Mechanics and Complex Systems at the University of Texas at Austin for helpful and stimulating discussions.

Bibliography

[1] Edward Nelson, *Dynamical Theories of Brownian Motion*, Princeton University Press, Princeton (1967); Edward Nelson, *Quantum Fluctuations*, Princeton University Press Princeton (1985). See also, Ph. Blanchard, Ph. Combe and W. Zheng, *Mathematical and Physical Aspects of Stochastic Mechanics*, Springer-Verlag, Heidelberg (1987), for further helpful discussion.

[2] For example, G.C. Ghirardi, P. Pearle and A. Rimini, *Phys. Rev.* A**42**, 78 (1990).

[3] L.P. Hughston, *Proc. Roy. Soc.* London A**452**, 953 (1996), and references cited there; S.L. Adler and L.P. Horwitz, *Jour. Math. Phys.* **41**, 2485 (2000).

[4] S.L. Adler, *Nuc. Phys.* B**415**, 195 (1994); S.L. Adler, *Quaternionic Quantum Mechanics and Quantum Fields*, Oxford Univ. Press, N.Y. (1995); S.L. Adler, hep-th/0206120.

[5] S.L. Adler and A.C. Millard, *Nuc. Phys.* B**473**, 199 (1996).

[6] G. Parisi, Y. Wu, Sci.Sin. **24** 483 ; Mikio Namiki *Stochastic Quantization* , (Springer-Verlag, Heidelberg, 1992).

[7] L.Morato and L.Viola, *J. Math. Phys.* **36** (1995) 4691-4710.

[8] E.C.G. Stueckelberg, *Helv. Phys. Acta* **14**, 322 (1941); **14**, 588 (1941).

[9] L.P. Horwitz and C. Piron, *Helv. Phys. Acta* **46**, 316 (1973). See also J. Schwinger, *Phys. Rev.* **82**, 664 (1951); R.P. Feynman, *Phys. Rev.* **80**, 440 (1950).

[10] See, for example, S.S. Schweber, *An Introduction to Relativistic Quantum Field Theory*, Harper and Row, N.Y. (1961).

[11] See for example, C. Itzykson and J.B. Zuber, *Quantum Field Theory* N.Y Mcgraw Hill, 1980.

[12] A.Einstein, *Ann. der Physik* **17** 1905 549-560. See also R. Furth. *Investigation of the Theory of Brownian Movement*, N.Y, Dutton (1956).

[13] L.P. Horwitz , O. Oron, *hep-ph/0205018.*

[14] S. Liberati, S. Sonego and M. Visser, *Ann. Phys.* 167 **298** (2002).

In: Relativity, Gravitation, Cosmology
Editor: V. Dvoeglazov and A. Espinoza Garrido, pp. 123-136
ISBN 1-59033-981-9

Chapter 10

A New Insight into the Negative-Mass Paradox of Gravity and the Accelerating Universe

Guang-Jiong Ni
Department of Physics, Fudan University, Shanghai, 200433, China
Department of Physics, Portland State University, Portland, OR 97207 USA
E-mail address: pdx01018@pdx.edu

Abstract

The discovery of acceleration of the universe expansion in recent astrophysics research prompts the author to propose that Newton's gravitation law can be generalized to accommodate the antimatter: While the force between matters (antimatters) is attractive, the force between matter and antimatter is a repulsive one. A paradox of negative-mass in gravity versus a basic symmetry ($m \to -m$) based on quantum mechanics is discussed in sufficient detail so that the new postulate could be established quite naturally. Corresponding modification of the theory of general relativity is also suggested. If we believe in the symmetry of particle and antiparticle as well as the antigravity between them, it might be possible to consider a new scenario of the expansion of universe which might provide some new insight into the interpretation of cosmological phenomena including the accelerating universe observed.

PACS: 95.30.Sf

10.1 Introduction

Let us begin with a "paradox of negative-mass" in the theory of gravity. As is well known, the Newton's gravitation law reads:

$$F(r) = -Gm_1m_2/r^2, \tag{10.1}$$

where m_1 or m_2 is the gravitational mass of a particle or a macroscopic body with spherical symmetry, r is the distance between them and G is the gravitational constant. The minus sign in (1) means that the force F between them is always an attractive one, which in turn implies that the gravitational potential energy of them is negative:

$$U(r) = -Gm_1m_2/r. \tag{10.2}$$

On the other hand, all experiments and the theory of special relativity (SR) have been verifying the equivalence of mass m and energy E, i.e., the Einstein's equation:

$$E = mc^2, \tag{10.3}$$

where m is the inertial mass. It is defined by another Newton's law of dynamics:

$$F = m\,a, \tag{10.4}$$

where a is the acceleration of the particle (body).

Consider a system (body) being composed of many particles, the gravitational binding energy shown by (2) would render the total mass of the body decreasing. Then an acute problem arises: can this body have a negative mass? If so, a very bizarre phenomenon would occur as discussed by Bondi, Schiff and Will respectively [1]: Suppose that such a body (with mass $m_1 < 0$) is brought close to a normal body (with mass $m_2 > 0$). According to Eqs. (1) and (4), the positive-mass body (m_2) would attract the negative-mass body(m_1) whereas the negative-mass body(m_1) would repel the positive-mass body (m_2). The pair (a "gravitational dipole") would accelerate itself off, without any outside help or use of propulsion! The conservation law of momentum and that of energy would all be violated.

Incredible! No one can believe in that. Hence we may name the above problem a "negative-mass paradox" in gravity. There are a number of paradoxes in physics, aiming at pushing the contradiction hidden in the theory to a situation that enables one to clarify what was wrong or missing in the basic concept.

To get rid of this "negative-mass paradox", a "positive-energy theorem" was posed in the middle of 1960s, saying that the total asymptotically determined mass of any isolated body in general relativity(GR) must be non-negative. This theorem had been proved since 1979 in a variety of ways and is in total conformity with Einstein's equation (3) because the observed inertial mass and energy are always positive definite.

However, being a paradox as acute as that hidden in Eq. (1), it does provide a much stronger hint than a negation of negative mass. Remembering that the appearance of negative energy is inevitable in relativistic quantum mechanics (RQM) and is intimately related to the existence of antiparticles, I will manage to claim that the whole solution to the "negative-mass paradox" must be a generalization of Newton's gravitation law into the following form:

$$F(r) = \pm G m_1 m_2 / r^2, \tag{10.5}$$

where the minus sign holds for m_1 and m_2 (both positive) being both matters or both antimatters whereas the plus sign holds for one of them being antimatter, which means that matter and antimatter repel each other.

10.2 Symmetry of space-time inversion

Starting from RQM, we consider the wavefunction (WF) of a freely moving (along x axis) particle:

$$\psi \sim exp\,[i(p\,x - E\,t)/\hbar], \tag{10.6}$$

where p is the momentum and $E(>0)$ the total energy.

But what is the WF ψ_c of an antiparticle? Before 1956, it was assumed to be a consequence of the operation of a so-called charge-conjugate transformation C which can bring a charged particle (say an electron with charge $-e$) to its antiparticle (say the positron with charge e) [2]:

$$\psi_c = C\psi \sim \psi^* \sim exp\,[i(-p\,x + E\,t)/\hbar]. \tag{10.7}$$

We see that the negative-energy $-E < 0$ emerges immediately due to the basic operators in quantum mechanics(QM):

$$\hat{p} = -i\hbar\frac{\partial}{\partial x}, \quad \hat{E} = i\hbar\frac{\partial}{\partial t}. \tag{10.8}$$

The negative-energy difficulty at the level of QM was remedied to some extent by the so-called "hole theory for positron" (which is obviously impossible for the Klein-Gordon particle) and solved formally at the level of quantum field theory (QFT) by the redefinition of creation (annihilation) operators.

Since the discovery of parity violation ,i.e., the violation of space-inversion $P(x \to -x)$ symmetry in 1956-1957, physicists realize that not only P but also C transformations are not conserved in the weak-interaction processes. So Eq. (7) is not applicable in general and the WF of antiparticle should be redefined as :

$$\psi_c = CPT\psi \sim exp\,[-i(p\,x - E\,t)/\hbar], \tag{10.9}$$

where the so-called "time-reversal transformation" T is defined as:

$$\psi \to T\psi = \psi^*(x, -t) \sim exp\,[i(-p\,x - E\,t)/\hbar]. \tag{10.10}$$

Note that: First, the name "time-reversal" is actually a misnomer[3, 4]. What the transformation (10) means is merely a reversal of motion ($p \to -p$). Second, the correctness of definition of antiparticle WF (9) depends on the validity of the CPT theorem which in turn is ensured by basic principles of SR and QFT. Third, as the complex-conjugate operations in C and T cancel each other, what a combined CPT transformation in (9) means is merely a sign change of coordinates (x, t) in comparison with Eq. (6) [2]. But the original meaning of C, P and T implies that Eq. (9) should describe an antiparticle with the same p and $E(>0)$ as that of the particle described by (6). Hence for antiparticles, we should forget the "hole theory" and use the following operators instead of (8):

$$\hat{p}_c = i\hbar\frac{\partial}{\partial x}, \quad \hat{E}_c = -\, i\,\hbar\frac{\partial}{\partial t}\,. \tag{10.11}$$

In fact, Eq. (11) had been proven to be the direct and unique outcome of the full solutions to the EPR paradox and Klein Paradox [5,4,6].

Fourth, once we accept Eqs.(9) and (11), the CPT theorem becomes a new fundamental postulate, i.e., a basic symmetry which can be stated in the following form:

Under the (newly defined) space-time inversion denoted by $\mathcal{PT}$, meaning merely $x \to -x$, $t \to -t$, the theory of RQM remains invariant with its concrete solution, e.g., a particle WF transforming to its antiparticle WF (denoted by $\mathcal{C}$) automatically. It means that our postulate reads:

$$\mathcal{PT} = \mathcal{C}. \tag{10.12}$$

For example, the Schrödinger equation is nonrelativistic. But the following coupled Schrödingerlike equation

$$\begin{cases} i\hbar\frac{\partial}{\partial t}\,\varphi = mc^2\varphi - \frac{\hbar^2}{2m}\,\nabla^2\,(\varphi+\chi), \\ i\hbar\frac{\partial}{\partial t}\,\chi = -mc^2\chi + \frac{\hbar^2}{2m}\,\nabla^2\,(\chi+\varphi), \end{cases} \tag{10.13}$$

is just the relativistic Klein-Gordon equation

$$(i\hbar\frac{\partial}{\partial t})^2\,\psi = -\,c^2\hbar^2\nabla^2\,\psi + m^2c^4\psi, \tag{10.14}$$

with relation first pointed out by Feshbach and Villars in 1958:

$$\begin{cases} \varphi = (\psi + i\,\hbar\,\dot{\psi}/mc^2)/2, \\ \chi = (\psi - i\,\hbar\,\dot{\psi}/mc^2)/2. \end{cases} \tag{10.15}$$

Now we see that under the space-time inversion $(x \to -x, t \to -t)$ and the transformation:

$$\varphi(-x,-t) \to \chi(x,t), \quad \chi(-x,-t) \to \varphi(x,t), \tag{10.16}$$

the Eq. (13) does remain invariant while a particle WF (6) with $|\varphi| > |\chi|$ (due to $E > 0$, see (15)) turning to its antiparticle WF (9) with $|\chi_c| > |\varphi_c|$ (due to $E < 0, E_c = -E > 0$, see (16)).

10.3 Symmetry of mass inversion

Alternatively, we can restate the above basic symmetry in the following way: Under the mass inversion:

$$m \to -m, \quad \varphi(x,t) \to \chi(x,t), \quad \chi(x,t) \to \varphi(x,t), \tag{10.17}$$

the theory, e.g., Eq. (13), remains invariant. Although transformation (16) is equivalent to transformation (17), they share different advantages. The former is relevant to unobservable coordinates (x, t) and so is more essential in RQM and equivalent to even more abstract symmetry of i versus $-i$ (see Eq.(6) versus (9)), while the latter is relevant to observable mass m and so is easily to be generalized to the case of classical theory.

For example, as is well known, the motion equation for an electron is given by the Lorentz-force law:

$$\boldsymbol{F} = m\boldsymbol{a} = -e\,(\boldsymbol{E} + \boldsymbol{v} \times \boldsymbol{B}/c). \tag{10.18}$$

Before 1956, its counterpart for positron is obtained via a C transformation $(-e \to e)$, yielding:

$$\boldsymbol{F}_c = m\boldsymbol{a}_c = e\,(\boldsymbol{E} + \boldsymbol{v} \times \boldsymbol{B}/c), \tag{10.19}$$

which is still allowed since C is conserved in the electromagnetic interaction. However, we'd better use the following working rule: to deal with particle (matter) and antiparticle(antimatter) on an equal footing, a classical theory must be invariant under the mass-inversion transformation $m \to -m$. Using this rule to Eq. (18), we get (19) immediately. Note that: First, m is always positive. Second, being the external field, the electric-(magnetic) field strength $\boldsymbol{E}(\boldsymbol{B})$ undergoes no change in the transformation. Third, in RQM like (13) or (14), the motion equation for antiparticle is the same one as that for particle. This is because each particle state like (6) contains its hidden antiparticle field under the condition $|\varphi| > |\chi|$ whereas an antiparticle state like (9) contains its hidden particle field under the condition $|\chi_c| > |\varphi_c|$. By contrast, in classical theory, the equation for particle is separated from that for antiparticle as shown by (18) versus (19). We just complement (18) with (19) so that the theory becomes invariant under the mass inversion transformation. In other words, a new equation must be added before it can be complete in the sense of treating the particle and antiparticle equally.

Fourth, to clarify further why we prefer the new postulate (12) instead of C transformation and CPT theorem, we wish to emphasize an important difference between a postulate (law) and a theorem. All quantities in a theorem must be defined in advance separately and unambiguously and the outcome of the theorem is actually contained in its premise implicitly. For example, the definitions of C, P and T are all clear in mathematics and the validity of CPT theorem is ensured by the basic principles of SR and QFT. Once C, P and T are not conserved in experiments, they cease to be meaningful as physical transformations. In this situation, the CPT theorem immediately exhibits itself as a new postulate (12) in which the definition of transformation of particle to antiparticle is just contained in the same equation. In general, a postulate or law can often (not always) accommodate a definition of physical quantity, and the validity of the postulate (law) together with the definition must be verified by experiments. Hence the establishment of a law is a process "from particular to general" or an outcome of "analysis and induction method". By contrast, to prove a theorem from well-established theories is a process "from general to particular" or a consequence of "deduction method".

For example, the definition of gravitational mass m is contained in the gravitational law, Eq. (1). The definition of inertial mass m is contained in Newton's law , Eq. (4). The definition of electric-magnetic field strength, $\boldsymbol{E}$ and $\boldsymbol{B}$, is contained in the Lorentz force law, Eq. (18). What we have done is a similar thing — the definition of particle-antiparticle transformation $\mathcal{C}$ is contained naturally in a new postulate (12) — not one that comes from elsewhere.

10.4 Generalization of Newton's gravitation law to antimatter and antigravity

We are now in a position to establish Eq. (5), performing the rule of mass-inversion on Eq. (1) together with Eq. (4). Note that, however, both m_1 and m_2 belong to a closed system under consideration. When one of them, say m_1, is an antimatter, we just perform a transformation $m_1 \to -m_1$ once, regardless of whether it is a source (m_1 appears only at one side of equation) or a detected body (m_1 appears at both sides of equation). This simple procedure brings Eq. (1) to Eq. (5) with a plus sign, implying that m_1 and m_2 repel each other. Furthermore, if m_2 is also an antimatter, one more transformation $m_2 \to -m_2$ recovers the minus sign, implying that two antimatters attract each other. Now the whole Eq. (5) is invariant with respect to transformation $m_i \to -m_i (i = 1, 2)$.

The reason we can get rid of the "negative-energy paradox" is as follows. If one thought that the detected body m_1 is really a negative mass ($m_1 < 0$ while $m_2 > 0$) seriously, the paradox was inevitable. But here every mass is positive regardless of whether it is matter or antimatter. Even if we make a careless mistake of performing transformation $m_1 \to -m_1$ twice, it is still not a serious mistake—we merely do a trivial thing to return back to the minus sign. Yet by another transformation on either m_1 or m_2 will the other equation with plus sign be supplemented. Eventually, the correct interpretation of whole Eq. (5) is independent of the procedures used to reach it. The bizarre phenomenon which occurred in the paradox will never happen again.

In some sense, the paradox stemmed from the incorrect conception that the difference between positive and negative masses is absolute, whereas now we regard the difference between m and $-m$ merely relative. We merely perform a symmetry transformation $m \to -m$ to show the equal existence of matter versus antimatter, but eventually there is no negative mass at all. This reminds us of the experience (i.e., lesson) of Einstein. He struggled for eight years before inventing the theory of GR. What Einstein eventually realized is that space and time have no absolute meaning but systems of relations (as stressed by Smolin [7]).

10.5 Generalization of Einstein field equation in general relativity

Consider a positronium and an atom of matter. If Eq. (5) is correct, there will be no gravitational force between them. This means that the gravitational mass m(grav.) of positronium is zero! However, the energy or the relevant inertial mass m(inert.) (see Eq. (3)) of positronium is definitely nonzero. Hence we see that in the case of coexistence of particles and antiparticles, the equivalence principle (EP) in the (weak) sense that [8]

$$m\,(\text{grav.}) = m\,(\text{inert.}) \tag{10.20}$$

cannot be valid.

As is well known, the EP served as a starting point in establishing the theory of general relativity(GR). The possible invalidity of EP in the presence of antimatter implies that GR is dealing with the gravitation of pure matter without the coexistence of antimatter. Indeed, let us look at the Einstein field equation [9]:

$$R_{\mu\nu} - \frac{1}{2} g_{\mu\nu} R = -8\pi G T_{\mu\nu} \tag{10.21}$$

On the left side, the Ricci tensor $R_{\mu\nu}$, curvature scalar R and the metric tensor $g_{\mu\nu}$ are all functions of coordinates x_μ. While on the right side, the energy-momentum tensor $T_{\mu\nu}$ is introduced to describe the existence of matter in the vicinity (a marcroscopic small volume) of x_μ. Then under a transformation of mass inversion $m \to -m$ to reflect that of matter to antimatter, $T_{\mu\nu}$ should change its sign due to its proportionality to the mass m. Hence Eq. (21) changes sign on the right side whereas not on the left side. This reflects the fact that GR is a classical field theory and so cannot treat the matter and antimatter on an equal footing.

To keep Eq. (21) invariant under the mass inversion, we manage to modify its right side by a generalization as:

$$T_{\mu\nu} \to T_{\mu\nu}^{eff} = T_{\mu\nu} - T_{\mu\nu}^{c}, \tag{10.22}$$

(where the superscript c means antimatter,) since under the mass inversion, $T_{\mu\nu} \to -T_{\mu\nu}^{c}$ and $T_{\mu\nu}^{c} \to -T_{\mu\nu}$. Notice that the form of the energy-momentum tensor is the same for both matter and antimatter. We stress once again that the distinction between m and $-m$ is merely relative, not absolute.

10.6 Antigravity and cosmology

Evidently, Eq. (5) cannot be tested by experiments on earth. We can only await and see new developments in astrophysical research. Surprisingly, in recent years, the careful observation of some distant Type Ia supernovae with redshift $z = 0.39 - 0.9$ reveals that the expansion of the universe is accelerating rather than decelerating as physicists pondered several years ago (see the excellent papers [10,11,12]). Many physicists tend to explain the acceleration by resorting to the cosmological constant (with dimension (length)$^{-2}$) originally introduced by Einstein, unsuccessfully trying to interpret it as a type of "vacuum zero-point energy" in QFT. It is often called "dark energy", which may account for up to 70% of the entire matter and energy in the universe. Actually, nobody understands what the dark energy really is.

Although I have no research experience in the field of GR and astrophysics, I do share the opinion of Davies [13] that "if we want to keep inflation and account for today's accelerating expansion, we need a theory that explains why antigravity was once tense, then dropped and hovered at just above zero,.... One possibility is that the force fades with time. Another is that it varies in space, so that far beyond the limit of our telescopes it may be much bigger,.... What we need is a theory that derives the strength of the antigravity force as part of a unified description of all the forces of nature." I am trying to pose some conjectures as follows.

(a). During the inflation era right after the big bang, once particles and antiparticles were created, they began to repel each other according to Eq. (5). I don't believe the interpretation that the present baryon asymmetry stemmed from the small CP violation (which is equivalent to T violation, as discussed above, it has nothing to do with the basic symmetry in time or that in particle and antiparticle). Rather, I prefer to think that the total number of particles is equal to that of antiparticles in the entire expanding universe. But antiparticles had a head start during the inflationary expansion and flew away faster than particles did. The latter lagged behind to some extent and our galaxies gradually evolved out of them. On the other hand, some distant stellar objects (quasars?) might be evolved from antiparticles and some of them may be just those observed Type Ia supernovae undergoing acceleration caused by the repulsive force exerted by inner galaxies composed mainly of matters.

(b). Based on the above picture and solving Eq. (21) with modification (22) in the Friedmann model as discussed in [9], we can see that in a large part of universe where the densities of matter and antimatter are nearly equal, the acceleration of the expansion tends to zero. In other words, the flatness of our universe as shown by the recent careful measurement of the fluctuation of cosmic microwave background radiation (see [14]) is critically depending on two factors: first, the inflationary expansion triggered by the repulsive force between matter and antimatter; second, the nearly equal densities of matter and antimatter in a large part of universe after the inflationary expansion. So a careful measurement of the variation of the Hubble constant [11] with distance may reveal the subtle difference in different regions.

(c). Among all particles, neutrinos are of particular interest [15]. After the big bang, superluminal neutrinos and antineutrinos are equally created in three flavors and distributed everywhere isotropically. Their number density may be 10^8 times that of protons. However, because the forces between neutrinos and matter just cancel that between antineutrinos and matter, all neutrinos and antineutrinos (even they all have positive tachyon mass, see VII,(d)) escape from the analysis of so-called dark matter.

(d). As the next conjecture, I guess that in the intermediate region where matter and antimatter are overlapping, there may be some probability of collisions between matter and antimatter. Because of the long range repulsive force between them, these collisions may be grazing and so the annihilation process occurs at short distance would have some special character. Is it possible to relate the above conjecture with the mechanism of gamma-ray-bursts(GRBs)? It has been noticed for years that GRBs are distributed at a remote distance on the cosmological scale and they tend to have a roughly constant explosive energy (see [16]and references therein). So the careful study of the GRB-Hubble diagram as conducted by Schaefer[16] is very important because the spatial distribution of GRB may provide further evidence of nonuniform distribution of matter and antimatter in the universe.

(e). Furthermore, I think that the properties of cosmic matter composed of pure matter and that of a mixture of matter with antimatter are certainly different. The latter bears some resemblance to plasma (not quite the same) with a "screening length". The fluctuations in plasma with length scale larger than the screening length should be strongly suppressed. In fact, the star formation rate is rising steadily from remote distance with redshift $z \sim 2$ to $z > 6$ [16] which seems to me a reflection that the domination of antimatter is increasing steadily there.

10.7 Summary and discussion

(a). I was pleasantly surprised to read the wonderful paper by Schödel et al. [17]. Their delicate observation over ten years and convincing analysis verified the existence of a huge black hole at the center of the Milky way. This should be viewed as a triumph of experimental science, also a triumph of the theory of GR , which was first verified by its successful explanation on the precession of Mercury's perihelion. However, despite all of these, GR is still a classical theory dealing with the gravitation of pure matter as shown by its noninvariance under the transformation of mass inversion.

(b). Based on the rule of mass-inversion invariance, the Newton's gravitation law is generalized to Eq. (5) and the Einstein field equation in GR, Eq. (21), is also modified by (22) such that they can treat matter and antimatter equally and explicitly. The antigravity between matter and antimatter may be important to account for many cosmological phenomena including the acceleration of expansion of the universe.

(c). Eq. (5) should be derived from Eq. (21) with (22) in a weak-field approximation. We follow the standard procedure as discussed in [9] and notice that in the case of one of the bodies being antimatter (the other matter), an extra minus sign will appear in front of the product of their energy-momentum tensors due to Eq. (22). This means antigravity in Eq. (5).

(d). We are dealing with three kinds of symmetries, the particle-antiparticle symmetry, the space-time inversion symmetry, and the mass inversion symmetry. They are essentially equivalent to each other. This can also be seen from the two Casimir invariants in the Poincaré group, which is a group generated by the inhomogeneous Lorentz transformation [18]:

$$x^{\mu} = \Lambda^{\mu}_{\nu} x^{\nu} + a^{\mu} \tag{10.23}$$

Usually, the conditions: $\det|\Lambda^{\mu}_{\nu}| = 1$ and $\Lambda^{0}_{0} \geq 1$ are imposed. So one is dealing with a proper and orthochronous Poincaré transformation. There are two Casimir invariants [commuting with all ten generators $M_{\mu\nu}$ (the six generators of the Lorentz group) and P_{μ} (the four generators of the space-time translation group)] being: $P_{\mu}P^{\mu}$ with eigenvalues m^2, (m is the particle mass) and $W_{\sigma}W^{\sigma}$ ($W_{\sigma} = -\frac{1}{2}\epsilon_{\mu\nu\rho\sigma}M^{\mu\nu}P^{\rho}$, the Pauli-Lubanski operator) with eigenvalues $-m^2 s(s+1)$, (s is the particle spin). Now we know that the physical essence of the Lorentz group is that of SR. The latter can be derived from any one of the three discrete symmetries mentioned above based on QM. Hence we can generalize the Poincaré group beyond the proper and orthochronous one to incorporate the antiparticle explicitly, i.e., we may perform the inversion of P_{μ} and $M^{\mu\nu}$ so that $P_{\mu} \to -P_{\mu}$, $W_{\sigma} \to -W_{\sigma}$ which means the transformation of a particle to its antiparticle. The invariance of $P_{\mu}P^{\mu}$ and $W_{\sigma}W^{\sigma}$ implies that their eigenvalues remain invariant under the corresponding inversion of mass: $m \to -m$. In other words, either the Lorentz group or the Poincaré group is relativistic in essence which is nothing but the equal existence of particle and antiparticle. Moreover, a negative eigenvalue of $P_{\mu}P^{\mu} = m^2 < 0$ is also allowed. This implies an analytical continuation of mass: $m \to im_s$ or $-im_s$ with m_s real (the tachyon mass) to describe a superluminal neutrino or antineutrino [15]. Here the symmetry between i and -i reflects even more deeply the symmetry of mass inversion which is really a relative one rather than an absolute one.

(e). The "photon" with $m^2 = 0$ in the representation of Poincaré group is by no means an ordinary particle. Usually, a photon with linear polarization (γ_x or γ_y) can be viewed as a linear superposition of a photon with left-handed polarization (γ_L) and that with right-handed polarization (γ_R). But in our unified theoretical scheme of space-time inversion, γ_L and γ_R could be viewed as the "particle" and "antiparticle" (or vice versa). So I guess that a photon has no gravitational mass. Alternatively, we may say that a photon exerts no gravitational force on other matter (antimatter). It just moves along the geodesic line in the space-time which is determined by all matter and antimatter in the entire universe. By the way, according to our understanding of so-called "wave-particle duality"[4], a "photon" should be treated theoretically as an electromagnetic wave in propagation until detected then it appears as a "particle". To speak of a photon's position before it is detected is meaningless.

(f). The possible violation of the equivalence principle in the case of coexistence of matter and antimatter has far-reaching implication. The energy or inertial mass [linked by Eq. (3)] obeys the conservation law and thus is observable. By contrast, the gravitational mass is not conserved and thus is not observable in the strict sense. It merely obtains its definition from Newton's gravitation law, which, of course, is a classical and approximate one.

(g). I dare not discuss the quantization of GR because I have little knowledge about the theory of quantum gravity (see the excellent introduction by Smolin [7]). However, according to our understanding about SR based on QM [4, 6], a possible "fourth road" to quantize GR might be a localization of the space-time inversion symmetry mentioned above (just like GR being the localization of the Lorentz symmetry in SR) so that both matter and antimatter can be treated explicitly and equally. Regrettably, I have been pondering this problem for many years with nearly no progress. So I hope that more physicists will join this difficult but interesting pursuit.

Acknowledgments

I thank S. Q. Chen, P. Leung, D. Lu and Z. Q. Shi for helpful discussions and bringing relevant references to my attention. I am greatly indebted to three referees for their kind comments leading to much improvement on this paper.

Bibliography

[1] Bondi, H. Negative mass in general relativity, *Rev. Mod. Phys.* 29, 423-428 (1957); Schiff, L. I. Sign of the gravitational mass of a positron, *Phys. Rev. Lett.* 1, 254- 255(1958); Will, C., *The renaissance of general relativity, in The New Physics*, Ch.2 (Edited by Davies, P., Cambridge Univ. Press, 1989)p.31.

[2] Bjorken, J. D. and Drell, S. D., *Relativistic Quantum Mechanics* (Mcgraw-Hill Book Company, 1964).

[3] Sakurai, J. J., *Modern Quantum Mechanics* (Revised Ed.)(John Wiley & Sons, Inc. 1994).

[4] Ni, G. J. and Chen, S. Q., *Advanced Quantum Mechanics* (Chinese Ed., Press of Fudan University, 2000, 2003), (English Ed., Rinton Press, 2002).

[5] Ni, G. J., Guan, H., Zhou, W. M. and Yan, J., Antiparticle in the light of Einstein-Podolsky-Rosen paradox and Klien paradox, *Chin. Phys. Lett.* 17, 393-395(2000), quant-ph/0001016.

[6] Ni, G. J., Ten arguments for the essence of special relativity, in *Proceedings of the 23rd Workshop on High-energy Physics and Field Theory*, (Protvino, Russia, June 2000, Edit: Filimonova, I. V. and Petrov, V. A.) 275-292; hep-th/0206250. An modified version will appear in *Progress of Physics* (Nanjing, China) 23(4)484-503(2003).

[7] Smolin, L., *Three Roads to Quantum Gravity* (Basic Books, 2001) p.149.

[8] Will, C. W., *Theory and Experiment in Gravitational Physics*, Revised Edition (Cambridge Univ. Press, 1993).

[9] Weinberg, S., *Gravitation and Cosmology* (John Wiley, 1972).

[10] Bahcall, N. A., Ostriker, J. P., Perimutter, S. and Steinhardt, P. J., The cosmic triangle: revealing the state of the universe, *Science*, 284, 28 May, 1481-1488(1999).

[11] Freedman, W., The Hubble constant and the expanding universe, *American Scientist*, 91, Jan-Fe. 36-43(2003).

[12] Livio, M., Moving right along, *Astronomy*, July, 34-39(2002).

[13] Davies, P., Seven wonders, *New Scientist*, 21 Sept. 28-33(2002).

[14] *New Scientist*, 2003, 15 Feb., 12-13;2003, 5 April, 22-23.

[15] Ni, G. J., A minimal three-flavor model for neutrino oscillation based on superluminal property, hep-ph/0306028, Chapter 11 of this book.

[16] Schaefer, B. E., Gamma-ray burst Hubble diagram to z=4.5, *The Astrophysical Journal*, 583: L67-L70(2003).

[17] Schödel, R., et al. A star in a 15.2-year orbit around the supermassive black hole at the centre of the Milky Way, *Nature*, 419, 17 Oct. 694-696 (2002).

[18] Marshak, R. E., *Conceptual foundations of modern particle physics* (World Scientific, 1993).

In: Relativity, Gravitation, Cosmology
Editor: V. Dvoeglazov and A. Espinoza Garrido, pp. 137-148
ISBN 1-59033-981-9

Chapter 11

A Minimal Three-flavor Model for Neutrino Oscillation based on Superluminal Property

Guang-Jiong Ni
Department of Physics, Fudan University, Shanghai, 200433, China
Department of Physics, Portland State University, Portland, OR 97207 USA
E-mail address: pdx01018@pdx.edu

Abstract

Recent experimental research progress of solar neutrino and neutrino oscillation reveals that neutrinos do have nonzero mass, which poses a serious challenge to theoretical physics. Two symmetries, the common invariance under the space-time inversion and the invariance or violation to maximum under the space inversion are analyzed for Dirac equation or the equation for superluminal neutrino respectively. Then a simple three-flavor model containing only one parameter is proposed and the relations among neutrino oscillation,parity violation and superluminal property are emphasized.

PACS: 14.60.Pq; 14.60.St

11.1 Introduction

Since the experimental verification of neutrino oscillation, i.e., the mutual transformation among three flavors of neutrinos (ν_e, ν_μ, and ν_τ) by Super-Kamiokande Collaboration [1] in 1998 and further by SNO Collaboration in 2002 [2], a common understanding that neutrinos have nonzero mass has been accepted by the physics community. This is because for a two-neutrino oscillation model, the probability for a neutrino produced in flavor state a to be observed in flavor state b after travelling a distance L through a vacuum is [1]:

$$P(\nu_a \to \nu_b) = \sin^2 2\theta \sin^2 \left(\frac{1.27 \triangle m^2 (\text{eV}^2) L(\text{km})}{E_\nu (\text{GeV})} \right) \tag{11.1}$$

where E_ν is the neutrino energy (in GeV), θ is the mixing angle between the flavor eigenstates and the mass eigenstates, and $\triangle m^2$ is the mass square difference of the neutrino eigenstates. Obviously, once $\triangle m^2 = 0$, there will be no neutrino oscillation at all (see the excellent books [3, 4]).

However, since the discovery of parity violation by Lee-Yang, Wu et al. in 1956-57 and the ingenious experiment by Goldhaber et al [5], it is concluded that neutrinos are always left-handed (ν_L) while antineutrinos are right-handed ($\overline{\nu}_R$). Once they have mass and so are Dirac particles moving at a speed u less than light-speed c, then an observer moving at a parallel velocity $v > u$ would see a neutrino becoming right-handed (ν_R) and an antineutrino left-handed ($\overline{\nu}_L$). But neither ν_R nor $\overline{\nu}_L$ exists in experiments. Hence the evidence for neutrino mass poses a serious challenge to theoretical physics [6].

There are some approaches to make neutrinos massive [3, 4]. One approach assumes that ν_R interacts much more weakly than other particles and so is unobserved. The second way involves a Majorana neutrino theory (in which a neutrino and antineutrino can be identical) in combination with the so-called seesaw mechanism which allows the extremely heavy ν_R to exist for a very short time interval and only allows ν_L to be observed. I have serious doubts about the above approaches, based on the criterion of beauty. As Dirac stressed: "Physical laws should have mathematical beauty" (see [7] foreword xiii and p.257). "It seems to be one of the fundamental features of nature that fundamental physical laws are described in terms of great beauty and power. As time goes on it becomes increasingly evident that the rules that the mathematician finds interesting are the same that Nature has chosen" (quoted from [8]). It seems to me that there is another very simple and interesting approach which makes neutrinos massive while in the meantime respecting parity violation, i.e., the permanently longitudinal polarization of neutrinos. Based on a series of works [9-15], I will discuss the equation and properties of a subluminal or superluminal particle before a simple three-flavor model for neutrino oscillation can be proposed.

11.2 Two symmetries of Dirac equation

Beginning from the famous Dirac equation, I wish to emphasize the essential beauty which is hidden in two symmetries. Let us write it in a form of two coupled equations of two-component spinors $\varphi_D(\boldsymbol{x}, t)$ and $\chi_D(\boldsymbol{x}, t)$, the hidden particle field and antiparticle field[16]:

$$\begin{cases} i\hbar\frac{\partial}{\partial t}\varphi_D = ic\hbar\boldsymbol{\sigma}\cdot\boldsymbol{\nabla}\chi_D + m_0c^2\varphi_D, \\ i\hbar\frac{\partial}{\partial t}\chi_D = ic\hbar\boldsymbol{\sigma}\cdot\boldsymbol{\nabla}\varphi_D - m_0c^2\chi_D, \end{cases} \tag{11.2}$$

Where $\boldsymbol{\sigma}$ are 2×2 Pauli matrices. The reason why we prefer this form rather than the covariant form of four-component spinor $\psi_D = \begin{pmatrix}\varphi_D \\ \chi_D\end{pmatrix}$ lies in the fact that the symmetries will become explicit as follows.

First, Eq. (2) is invariant under the (newly defined) space-time inversion $(\boldsymbol{x}\to -\boldsymbol{x}, t\to -t)$with transformation (see [16]):

$$\varphi_D(-\boldsymbol{x},-t)\to\chi_D(\boldsymbol{x},t),\quad \chi_D(-\boldsymbol{x},-t)\to\varphi_D(\boldsymbol{x},t). \tag{11.3}$$

When $|\varphi_D|>|\chi_D|$, it is a particle; when $|\chi_D^c|>|\varphi_D^c|$, it shows up as an antiparticle. The superscript (or subscript) c refers to the antiparticle. Next, after we define the hidden left-handed and right-handed fields as:

$$\xi_D = \frac{1}{\sqrt{2}}(\varphi_D+\chi_D),\quad \eta_D = \frac{1}{\sqrt{2}}(\varphi_D-\chi_D). \tag{11.4}$$

Eq. (2) is recast into:

$$\begin{cases} i\hbar\frac{\partial}{\partial t}\xi_D = ic\hbar\boldsymbol{\sigma}\cdot\boldsymbol{\nabla}\xi_D + m_0c^2\eta_D, \\ i\hbar\frac{\partial}{\partial t}\eta_D = -ic\hbar\boldsymbol{\sigma}\cdot\boldsymbol{\nabla}\eta_D + m_0c^2\xi_D. \end{cases} \tag{11.5}$$

When $|\xi_D|>|\eta_D|$, it is a left-handed particle; when $|\eta_D|>|\xi_D|$, it becomes right-handed explicitly. Now we see that Eq. (5) is invariant under the pure space-inversion $(\boldsymbol{x}\to -\boldsymbol{x}, t\to t)$ with transformation:

$$\xi_D(-\boldsymbol{x},t)\to\eta_D(\boldsymbol{x},t),\quad \eta_D(-\boldsymbol{x},t)\to\xi_D(\boldsymbol{x},t). \tag{11.6}$$

The solution of either (2) or (5) gives the wave-function(WF) describing a freely moving Dirac particle:

$$\varphi_D\sim\chi_D\sim\xi_D\sim\eta_D\sim exp\left[i\left(\boldsymbol{p}\cdot\boldsymbol{x}-Et\right)/h\right], \tag{11.7}$$

with the momentum $\boldsymbol{p}$ and energy E satisfying the kinematical relation in special relativity (SR):

$$E^2 = p^2c^2 + m_0^2c^4. \tag{11.8}$$

The Dirac equation was discovered as an outcome of the combination of quantum mechanics (QM) with the theory of SR. However, what we described above should be viewed as a derivation of Eq.(8) in SR from QM by injecting two symmetries[16]. Just as Gross stressed: “the primary lesson of physics in 20th century is that the secret of nature is symmetry”.[8]

11.3 Parity violation and equation for superluminal particle

Based on Dirac's experience together with a belief that a particle is always impure without exception , we manage to establish the equation for a superluminal neutrino as:

$$\begin{cases} i\hbar\dfrac{\partial}{\partial t}\,\varphi = i c\hbar\boldsymbol{\sigma}\cdot\boldsymbol{\nabla}\chi + m_s\,c^2\,\chi, \\ i\hbar\dfrac{\partial}{\partial t}\,\chi = i c\hbar\boldsymbol{\sigma}\cdot\boldsymbol{\nabla}\varphi - m_s\,c^2\,\varphi, \end{cases} \tag{11.9}$$

or equivalently:

$$\begin{cases} i\hbar\dfrac{\partial}{\partial t}\,\xi = i c\hbar\boldsymbol{\sigma}\cdot\boldsymbol{\nabla}\xi - m_s\,c^2\,\eta, \\ i\hbar\dfrac{\partial}{\partial t}\,\eta = -i c\hbar\boldsymbol{\sigma}\cdot\boldsymbol{\nabla}\eta + m_s\,c^2\,\xi, \end{cases} \tag{11.10}$$

with Eq. (4) without subscript D. We see that while Eq. (9) remains invariant under the space-time inversion (3), Eq. (10) is violated (to maximum) under the space inversion (6) due to opposite signs in the mass term. A similar plane-wave solution like (7) now yields:

$$E^2 = p^2c^2 - m_s^2c^4, \tag{11.11}$$

which in turn implies a kinematical relation being:

$$p = \frac{m_s u}{\sqrt{\frac{u^2}{c^2}-1}}, \qquad E = \frac{m_s c^2}{\sqrt{\frac{u^2}{c^2}-1}}, \tag{11.12}$$

where the particle velocity u is defined as the group velocity u_g of wave ($E = \hbar\omega, p = \hbar k$):

$$u = u_g = \frac{d\omega}{dk} = \frac{dE}{dp} = \frac{c^2 p}{E} > c. \tag{11.13}$$

Hence, it is a superluminal particle (or tachyon) and m_s is called the proper mass or tachyon mass. On the other hand, the phase velocity u_p of wave reads:

$$u_p = \omega/k = E/p < c, \tag{11.14}$$

because Eq. (11) (like (8)) always imposes a constraint on the product of u_g with u_p:

$$u_g u_p = c^2. \tag{11.15}$$

Some consequences of the above theory are as follows:

(a) As discussed in Ref[16], a particle and its antiparticle obey the same equation, e.g., the Dirac Eq. (2) or the neutrino Eq. (9). The WF for a particle under the condition $|\varphi| > |\chi|$ has positive-energy $E > 0$ whereas that for its antiparticle under the condition $|\chi_c| > |\varphi_c|$ has negative-energy $E < 0$ (actually it has a positive-energy $E_c = -E > 0$, see [16]).

(b) In contrast to the Dirac Eq. (5), the neutrino Eq. (10) only allows solutions of ν_L (with $|\xi| > |\eta|$) and $\overline{\nu}_R$ (with $|\eta_c| > |\xi_c|$), the other two kinds of solution — ν_R and $\overline{\nu}_L$ — are strictly forbidden. This is a direct consequence of parity violation. Because now ν_L and $\overline{\nu}_R$ have velocity $u > c$, their polarizations (helicities) in one frame can be maintained in another frame with relative velocity $v(< c)$ between two observers.

(c) A comparison between Eq. (8) and (11) reveals that if $m_0^2 = -m_s^2$, we may write $m_0 = im_s$ (with real m_s). It has often been said that the kinematical relation for a superluminal particle could be viewed as some analytic continuation of that for a subluminal particle.

Experimentally, in measurements like that of tritium beta decay, physicists did find that in the kinematical relation of the neutrino

$$E^2 = p^2c^2 + m_\nu^2 c^4,$$

the mass square seems negative with large uncertainty:

$$\begin{aligned} m^2(\nu_e) &= -2.5 \pm 3.3 \text{ eV}^2, \\ m^2(\nu_\mu) &= -0.016 \pm 0.023 \text{ MeV}^2, \end{aligned} \tag{11.16}$$

as summarized in the particle table published in 2000 [17]. (See Eqs. (29) and (30).)

(d)While the Dirac Eq. (2) and the superluminal Eq. (9) share common basic symmetry under the space-time inversion (3), they have different symmetries under the space inversion (6): the Dirac Eq. (5) remains invariant whereas the superluminal Eq. (10) shows maximum violation, which should be viewed as some antisymmetry rather than asymmetry. In short, either the Dirac equation or the superluminal equation is destined for a specific realization of symmetries.

11.4 Neutrino oscillation–two flavor model

We are now in a position to discuss neutrino oscillation, generalizing Eq. (10) to a model for two flavors, ν_e and ν_μ. Similar to the form used in the discussion of coupled neutrinos with nonzero rest masses[3,4,18], now the coupled equation is assumed to be ($m_e = m_s(\nu_e), m_\mu = m_s(\nu_\mu), \hbar = c = 1$):

$$\begin{cases} i\,\dot{\xi}_e = i\boldsymbol{\sigma}\cdot\nabla\xi_e - m_e\eta_e - \delta\,\eta_\mu, \\ i\,\dot{\eta}_e = -i\boldsymbol{\sigma}\cdot\nabla\eta_e + m_e\xi_e + \delta\,\xi_\mu, \\ i\,\dot{\xi}_\mu = i\boldsymbol{\sigma}\cdot\nabla\xi_\mu - m_\mu\eta_\mu - \delta\,\eta_e, \\ i\,\dot{\eta}_\mu = -i\boldsymbol{\sigma}\cdot\nabla\eta_\mu + m_\mu\xi_\mu + \delta\,\xi_e\,. \end{cases} \tag{11.17}$$

Notice that: First, the additional terms with coupling constant δ do not spoil the symmetry in (10); Eq. (17) still remains invariant under space-time inversion

(3) and in the meantime violates space-inversion symmetry (6) to maximum. Second, like Eq. (10), Eq. (17) has the probability density being:

$$\rho = \xi_e^\dagger \xi_e + \xi_\mu^\dagger \xi_\mu - \eta_e^\dagger \eta_e - \eta_\mu^\dagger \eta_\mu, \tag{11.18}$$

which is not positive definite: While $\int \rho d\boldsymbol{x} = 1$ means the oscillation occurred between left-handed ν_e and ν_μ ($|\xi| > |\eta|$), $\int \rho d\boldsymbol{x} = -1$ means that between right-handed $\overline{\nu}_e$ and $\overline{\nu}_\mu$ ($|\eta_c| > |\xi_c|$).

Third, using the trick and notation similar to that of Sassaroli[18], we find two energy eigenvalues of Eq. (17) as:

$$\begin{aligned} &E_1^2 = p^2 - m_1^2, \quad E_2^2 = p^2 - m_2^2, \\ &m_{1,2} = \tfrac{1}{2}[(m_e + m_\mu) \pm R], \\ &R = \sqrt{(m_\mu - m_e)^2 + 4\delta^2}. \end{aligned} \tag{11.19}$$

Denoting $\Lambda = \dfrac{m_\mu - m_e + R}{2\delta}$, $\sin\theta = 1/\sqrt{1+\Lambda^2}$, $\cos\theta = \Lambda/\sqrt{1+\Lambda^2}$, $\nu_\mu = \begin{pmatrix} \xi_\mu \\ \eta_\mu \end{pmatrix}$, $\nu_e = \begin{pmatrix} \xi_e \\ \eta_e \end{pmatrix}$, we can express the mass eigenstates as (see Appendix):

$$\begin{pmatrix} \nu_1 \\ \nu_2 \end{pmatrix} = \begin{pmatrix} \cos\theta & \sin\theta \\ -\sin\theta & \cos\theta \end{pmatrix} \begin{pmatrix} \nu_\mu \\ \nu_e \end{pmatrix}. \tag{11.20}$$

Hence the time evolution of flavor amplitude is given by:

$$\begin{pmatrix} C_\mu(t) \\ C_e(t) \end{pmatrix} = \begin{pmatrix} \cos\theta & -\sin\theta \\ \sin\theta & \cos\theta \end{pmatrix} \begin{pmatrix} C_1(0)e^{-iE_1 t} \\ C_2(0)e^{-iE_2 t} \end{pmatrix}. \tag{11.21}$$

Suppose that at $t = 0$, an electron-neutrino is produced:

$$C_\mu(0) = 0, \quad C_e(0) = 1. \tag{11.22}$$

We obtain the oscillation probability of ν_e to ν_μ after a time t being:

$$P(\nu_e \to \nu_\mu) = |C_\mu(t)|^2 = \sin^2 2\theta \sin^2\left(\frac{(m_1^2 - m_2^2)t}{4E}\right), \tag{11.23}$$

while

$$|C_e(t)|^2 = 1 - \sin^2 2\theta \sin^2\left(\frac{(m_1^2 - m_2^2)t}{4E}\right), \tag{11.24}$$

where an ultrarelativistic approximation ($p \gg m, p \sim E$) has been used. Eq. (23) is exactly Eq. (1) ($t \sim L$) except that now $\triangle m^2 = m_1^2 - m_2^2$ refers to the square difference of tachyon mass for a superluminal neutrino rather than that of rest mass for a subluminal neutrino.

Since the experimental data show that the oscillation amplitude tends nearly to maximum, we would expect in the above two-flavor model that $\theta \to \pi/4$, or $\Lambda \to 1$, i.e., $\delta \gg (m_\mu - m_e)$. But a vanishing of m_μ and m_e is not allowed because it would eliminate the oscillation entirely.

11.5 Neutrino oscillation–a minimal three-flavor model

Interesting enough, we can keep only one coupling parameter in a three-flavor model by first generalizing (17) without the mass terms into:

$$\begin{cases} i\,\dot{\xi}_e = i\boldsymbol{\sigma}\cdot\boldsymbol{\nabla}\xi_e - \delta\eta_\mu - \varepsilon\,\eta_\tau, \\ i\,\dot{\eta}_e = -i\boldsymbol{\sigma}\cdot\boldsymbol{\nabla}\eta_e + \delta\xi_\mu + \varepsilon\,\xi_\tau, \\ i\,\dot{\xi}_\mu = i\boldsymbol{\sigma}\cdot\boldsymbol{\nabla}\xi_\mu - \lambda\eta_\tau - \delta\,\eta_e, \\ i\,\dot{\eta}_\mu = -i\boldsymbol{\sigma}\cdot\boldsymbol{\nabla}\eta_\mu + \lambda\xi_\tau + \delta\,\xi_e \\ i\,\dot{\xi}_\tau = i\boldsymbol{\sigma}\cdot\boldsymbol{\nabla}\xi_\tau - \varepsilon\eta_e - \lambda\,\eta_\mu, \\ i\,\dot{\eta}_\tau = -i\boldsymbol{\sigma}\cdot\boldsymbol{\nabla}\eta_\tau + \varepsilon\xi_e + \lambda\,\xi_\mu, \end{cases} \tag{11.25}$$

and then set $\delta = \varepsilon = \lambda$ later. By solving a typical algebraic equation of the third order, we find three eigenvalues of energy square being:

$$\begin{aligned} &E_i^2 = p^2 - m_i^2, \quad (i = 1, 2, 3) \\ &m_1^2 = 4\delta^2, \quad m_2^2 = m_3^2 = \delta^2\,. \end{aligned} \tag{11.26}$$

The orthogonal transformation between mass and flavor eigenstates reads:

$$\begin{pmatrix} \nu_1 \\ \nu_2 \\ \nu_3 \end{pmatrix} = \begin{pmatrix} \frac{1}{\sqrt{3}} & \frac{1}{\sqrt{3}} & \frac{1}{\sqrt{3}} \\ \frac{1}{\sqrt{6}} & \frac{1}{\sqrt{6}} & \frac{-2}{\sqrt{6}} \\ \frac{1}{\sqrt{2}} & \frac{-1}{\sqrt{2}} & 0 \end{pmatrix} \begin{pmatrix} \nu_e \\ \nu_\mu \\ \nu_\tau \end{pmatrix}. \tag{11.27}$$

(We should write $m_1 = 2\delta$ and $m_2 = m_3 = -\,\delta$ in the proof.) Similar to the procedure leading from (20) to (24), we get:

$$\begin{aligned} &|C_e(t)|^2 = 1 - \tfrac{8}{9}\sin^2[\tfrac{1}{2}(E_2 - E_1)t], \\ &|C_\mu(t)|^2 = |C_\tau(t)|^2 = \tfrac{4}{9}\sin^2[\tfrac{1}{2}(E_2 - E_1)t]\,. \end{aligned} \tag{11.28}$$

11.6 Summary and discussion

(a). If neutrinos are propagating through a medium, the oscillation picture will be modified drastically as discussed by Wolfenstein[19] and Mikheev-Smirnov[20]. The finite length of the wave packet must be considered[20, 3]. While different phase velocities (u_p) lead to oscillations, different group velocities (u_g) result in a separation of the wave packets in space. The situation becomes more complex because Eq. (15) now means $u_g > c$ in the superluminal case whereas $u_p < c$, implying that during propagation, the coherent superposition of different flavor states may be difficult to catch up with the wave packets which are separating faster than light. Hence I guess that the periodical flavor transformation probability (28) must be strongly suppressed in a medium. Maybe the oscillation becomes so incoherent that it actually provides a mechanism of detailed balancing for neutrinos. So eventually an equally distributed population may be established among the three flavors. Could the experimental data of solar neutrinos [2] be explained in this way? I can only place my hope on experts in this field.

(b). As stressed by Close[21], solving the solar neutrino problem qualifies as one of the great moments in experimental science and neutrino physics is about to enter a golden age. Now the SNO experiment[2] and the Kamland experiment in Japan have actually fixed three flavors of neutrino and finally ruled out the existence of a possible fourth flavor (or so called 'sterile' neutrinos.) Furthermore, by the K2K experiment in Japan[6] and some other neutrino factories under construction or planning, physicists will be able to detect neutrinos travelling long distances on Earth and measure their oscillations as well as their velocities. I was excited to learn the preliminary data of neutrino velocity $v = 3.0120481 \times 10^8$ m/s along a distance of 250 km by K2K in June 1999 [22]. So I am anxious to see the new data as soon as possible.

(c). If we measure the mass square m^2 of a neutrino with a certain flavor when it is just produced, like ν_e in (28), then m^2 must be an average of (mass)^2 eigenvalues with considerable standard deviation σ due to the existence of oscillation. For instance, in the above three-flavor model, we may predict a measured value of m^2 being:

$$m^2 = < m^2 > \pm\sigma = -\frac{8}{5}\,\delta^2 \pm \frac{6}{5}\,\delta^2\,. \tag{11.29}$$

But if the neutrino is an outcome of oscillation, like ν_μ or ν_τ in (28), then its mass square $\widetilde{m}^2$ measured in flight is expected to be:

$$\widetilde{m}^2 = < \widetilde{m}^2 > \pm\widetilde{\sigma} = -\frac{5}{2}\,\delta^2 \pm \frac{3}{2}\,\delta^2\,. \tag{11.30}$$

These two predictions are independent of the flavor in this model (Compare Eq. (16)).

(d). If we modify all coupling terms in (25) into plus signs, then instead of (26), we obtain:

$$\begin{aligned} &E_i^2 = p^2 + m_i^2, \quad (i = 1, 2, 3) \\ &m_1^2 = 4\delta^2, \quad m_2^2 = m_3^2 = \delta^2\,. \end{aligned} \tag{11.31}$$

However, similar to (28), Eq. (31) implies oscillation among three flavor states of subluminal Dirac particles. But in fact, there is no oscillation among the electron, the muon and the tau lepton. Hence we understand why the muon-electron transformation process like $\mu \to e + \gamma$ is forbidden. This is because the flavor change term is stemming from weak interactions with parity violation, if it is combined with the Dirac equation (of μ or e) which is parity conserved, it would make the whole equation asymmetric under space inversion. In other words, it is the inconsistency of symmetry that forbids the process of muon-electron transformation.

(e). Actually, our model is a natural extension of the standard model—the Weinberg-Salam theory, where each flavor of the massless neutrino obeys the Weyl equation and so is moving at the speed of light to maintain its permanent helicity. Once the couplings among the three flavors exist and oscillations occur, then Eq. (25) is inevitable and exhibits itself as the only motion equation shared by all three flavors of a superluminal neutrino or antineutrino. A coupling constant (δ) with the dimension of energy is the main fundamental hypothesis of our model, its validity should be justified by experiments in the years to come [24].

(f). In an another paper[23], it is argued that the Newton's gravitation law should be generalized to accommodate antimatter, e.g., while the gravitational force between neutrinos and matter is attractive, that between antineutrinos and matter is a repulsive one. Therefore, although neutrinos and antineutrinos in equal quantities (with number density up to 10^8 times that of protons) are distributed isotropically in the entire universe and they do have nonzero inertial mass, they show no gravitational mass as a whole. In other words, their existence violates the equivalence principle in the theory of general relativity. Moreover, this is why all neutrinos and antineutrinos escape from the analysis of so-called dark matter.

(g). In summary, there are three things being intimately correlated together: As Dirac particles obey the parity conservation law, they are bound to be subluminal and cannot oscillate among different flavor states. By contrast, since neutrinos violate the parity symmetry to maximum, they must be superluminal particles and are capable of oscillating among three flavors.

Acknowledgments

I thank S. Q. Chen, Z. X. Huang, P. Leung, D. Lu and Z. Q. Shi for bringing relevant references to my attention and helpful discussions. I am also grateful to referees for their kind comments leading to considerable improvement of this paper.

Appendix: Mass eigenstates of neutrino in two-flavor model

Assuming the neutrino momentum $\boldsymbol{p}$ along z axis and denoting $\xi = \begin{pmatrix} 0 \\ 1 \end{pmatrix}$ $\left[\eta = \begin{pmatrix} 1 \\ 0 \end{pmatrix}\right]$ the left-handed (right-handed) polarized spin state, we can express two normalized solutions of Eq. (17) for neutrino (with energy $E_{1,2} > 0$) being:

$$\psi_1(z,t) = \frac{1}{\sqrt{V}}\theta_1(p)e^{i(pz-E_1t)}, \quad \theta_1(p) = \frac{1}{\sqrt{1+\Lambda^2}}\begin{pmatrix} u_1(p) \\ \Lambda u_1(p) \end{pmatrix},$$
$$u_1(p) = \sqrt{\frac{p+E_1}{2E_1}}\begin{pmatrix} \xi \\ \frac{m_1}{p+E_1}\xi \end{pmatrix}, \tag{11.32}$$

$$\psi_2(z,t) = \frac{1}{\sqrt{V}}\theta_2(p)e^{i(pz-E_2t)}, \quad \theta_2(p) = \frac{1}{\sqrt{1+\Lambda^2}}\begin{pmatrix} \Lambda u_2(p) \\ -u_2(p) \end{pmatrix},$$
$$u_2(p) = \sqrt{\frac{p+E_2}{2E_2}}\begin{pmatrix} \xi \\ \frac{m_2}{p+E_2}\xi \end{pmatrix}. \tag{11.33}$$

This is Eq. (20). The WFs for antineutrino with positive energy $E_c = -E, (E = -E_{1,2})$ are:

$$\psi_{c1}(z,t) = \frac{1}{\sqrt{V}}\theta_3(p)e^{-i(pz-E_1t)}, \quad \theta_3(p) = \frac{1}{\sqrt{1+\Lambda^2}}\begin{pmatrix} u_3(p) \\ \Lambda u_3(p) \end{pmatrix},$$
$$u_3(p) = \sqrt{\frac{p-E_1}{2E_1}}\begin{pmatrix} \eta \\ \frac{m_1}{p-E_1}\eta \end{pmatrix}, \tag{11.34}$$

$$\psi_{c2}(z,t) = \frac{1}{\sqrt{V}}\theta_4(p)e^{-i(pz-E_2t)}, \quad \theta_4(p) = \frac{1}{\sqrt{1+\Lambda^2}}\begin{pmatrix} \Lambda u_4(p) \\ -u_4(p) \end{pmatrix},$$
$$u_4(p) = \sqrt{\frac{p-E_2}{2E_2}}\begin{pmatrix} \eta \\ \frac{m_2}{p-E_2}\eta \end{pmatrix}. \tag{11.35}$$

Bibliography

[1] Y. Fukuda, *et al.* Evidence for oscillation of atmospheric neutrinos, *Phys. Rev. Lett.* 81, 1562-1567(1998).

[2] Q. R. Ahmad, *et al.* Direct evidence for neutrino flavor transformation from neutral-current interactions in the Sudbury Neutrino Observatory. *Phys. Rev. Lett.* 89, 011301-1-6(2002).

[3] C. W. Kim and A. Pevsner, *Neutrino in Physics and Astrophysics*, (Harwood Academic Publishers, 1993).

[4] R. N. Mohapatra and P. B. Pal, *Massive Neutrinos in Physics and Astrophysics* (World Scientific, 1991).

[5] M. Goldhaber, L. Grozins and A. Sunyar, Helicity of neutrinos, *Phys. Rev.* 109, 1015-1017(1958).

[6] H. Murayama, The origin of neutrino mass, *Phys. World*, May 2002, 35-39.

[7] G. Farmelo, Edited, *It Must Be Beautiful–Great Equations of Modern Science*, (Granta Books, London, New York, 2002).

[8] D. J. Gross, *Gauge theory–past, present and future*, talk delivered at International Symposium in Honor of C. N. Yang's 70th Birthday, National Tsing Hua University, Hsinchu, Taiwan, July 1992.

[9] A. Chodos, A. I. Hauser and V. A. Kostelecky, The neutrino as a tachyon, *Phys. Lett.* B150, 431-435(1985).

[10] E. Giannetto, G. D. Maccarrone, R. Mignani and E. Recami, *Are muon neutrinos faster-than light particles? ibid*, 178, 115-119(1986).

[11] J. Ciborowski and J. Rembielinski, Tritium decay and the hypothesis of tachyonic neutrinos, *Europ. Phys. J.* C8, 157-161(1999).

[12] T. Chang and G. J. Ni, An explanation of possible negative mass-square of neutrino, *Fizika* B (Zagreb)11, 49-55(2002), hep-ph/0009291.

[13] G. J. Ni and T. Chang, Two parameters describing a superluminal neutrino, *Jour. of Shaanxi Normal University (Nat. Sci. Edi.)* 30(3) 32-39(2002), hep-ph/0103051.

[14] G. J. Ni, *There might be superluminal particles in nature*, *ibid*, 29(3) 1-6(2001), hep-th/0201077; Superluminal paradox and neutrino, *ibid*, 30(4) 1-6(2002), hep-ph/0203060.

[15] G. J. Ni, Evidence for neutrino being likely a superluminal particle, submitted to *Int. J. Mod. Phys.* A, hep-ph/0206296.

[16] G. J. Ni and S. Q. Chen, *Advanced Quantum Mechanics*, (Chinese Ed.: Press of Fudan University, 2000, 2003); (English Ed.: Rinton Press, 2002).

[17] Particle Data Group, Review of Particle Physics, *Euro. Phys. J.* C15, 350(2000).

[18] E. Sassaroli, Neutrino oscillation: a relativistic example of a two-level system, *Am. J. Phys.* 67, 869-875(1999).

[19] L. Wolfenstein, Neutrino oscillations in matter, *Phys. Rev.* D17, 2369-2374(1978).

[20] S. P. Mikheev and A. Y. Smirnov, Resonance oscillations of neutrinos in matter, *Soviet Phys.-USPEKHI* 30, 759-790(1987).

[21] F. Close, To catch a rising star, Reluctant heroes, *New Scientist*, 7 Dec. 33-36(2002).

[22] Z. X. Huang, Modern advances in neutrino researches, *Engineering Science* (in Chinese) 4(10), 90-93(2002).

[23] G. J. Ni, A new insight into the negative-mass paradox of gravity and the accelerating universe, *physics/0308038, Chapter 10 of this book*, preprint.

[24] G. J. Ni, Cosmic ray spectrum and tachyonic neutrino, *hep-ph/0404030.*

In: Relativity, Gravitation, Cosmology
Editor: V. Dvoeglazov and A. Espinoza Garrido, pp. 149-161
ISBN 1-59033-981-9

Chapter 12

The World of Events Reality: Instantaneous Action as a Connection of Events through Time

I. A. Eganova
Sobolev Institute of Mathematics SD RAS, Novosibirsk, 630090, Russia

Abstract

N. A. Kozyrev's point of view is presented in the light of the World of events (space-time) reality: instantaneous action is the appearance of the innate connection of events with the null proper time interval, this phenomenon is conditioned by the temporal aspect of the World of events. In the paper the main properties of the phenomenon and its physical origin are considered and the corresponding experimental results are given.

"What we are talking about... could be done long ago, after the work by Minkowski, and certainly he himself would explain just this all if he had not been gone so early. However, the dogmatism and faith — they are always alien to the science and constantly accompanying it — did their work. Nearly to our time they restricted the level of the understanding and, as consequence, narrowed the field of the relativity application."

/A. A. Logunov [1, p. 3]/

12.1 Introduction

97 years ago in physics appeared the four-dimensional object (the 'event' — one of the key concepts of the modern theoretical physics) in well-known paper by H. Poincaré [2], see also [3]. Two years later H. Minkowski [4] presented the four-dimensional mathematical model of the physical reality (the 'World' ('Universe'), 'space-time') as a completely new field of physics, see Logunov's lectures [1], where it was analysed in detail.

The fundamental nature of the hypothesis on the four-dimensionality of the physical reality, on united pseudo-Euclidean geometry of space and time — that is *quinta essentia* of the theory of relativity, see [1] — is emphasized in many works devoted to the space-time, e.g., [5]–[7]. J. L. Synge [6],[7] considers this hypothesis as the most fundamental one out of all which were ever advanced. It discovered the new real world — World of events. A. Einstein in his lectures on the essence of the theory of relativity [8] which he read at the Princeton University in May 1921 accentuated that neither a point of space nor the moment of time, when something happened, possess the physical reality, but only the event itself does. "We deal with *events*. So has physics from its beginning. It requires *four* instrument readings (never mind what instruments) to specify an event, and so the totality of events forms a 4-dimensional continuum — we call it *space-time*. It is a single concept, not a combination of the separate concepts of *space* and *time*" [9, p. 34]. In other words, the essence of the World of events consists in deep interconnection of the spatial and temporal aspects of the material World which manifests itself in the universally adopted mathematical truth that the space-time in the general case cannot be separated into the space and time in some invariant way [7].

If in space-time physics we proceed actually from the notions of the space and time as two opposite, mutually complementing structures of existence of the material World, we have to expect that the proper study of the physical properties of its temporal structure, the reality of the World of events, would unveil new possibilities and perspectives, essentially new types of interconnection of phenomena and essentially new methods to change the condition of substance. That is why we have to study the physical properties of the temporal aspect of natural systems, phenomena and processes as well as of the spatial one. Note, in particular, the four-dimensional model of the physical reality called attention to the definition of time. A. A. Friedman in his monograph [5] began to develop some mathematical approach to the problem of time in order "to return to time in physics its exceptional position" connected with causality. Unfortunately, this approach is not realized: soon after his author is gone.

N. A. Kozyrev was the first who in accordance with the essence of the World of events, on the one hand, but, on the other hand, in search of ways of tackling the fundamental astronomical problem of the nature of the stellar energy [10, pp. 71–154, 191–204] realized the direct investigation (1948–1983) of the physical properties of time, its part in the fundamental laws of Universe existence: in his concept of time N. A. Kozyrev was able to overcome I. Newton's formal identification of the 'time' with the 'duration' (see Newton's definition of the "absolute, true, mathematical time" in [11, p. 6]), which was predominant in physics.

In 1959 J. L. Synge proposed that the word *"chronometry"* ought to be used for denotement of that part of science which deals with the concept of time, with the same wide scope which we have learned to give to the word *"geometry"* [12]. His book [7] was the first capital work in the field of the theoretical chronometry.

At present theoretical investigations which are based on the view-point of the conception of the four-dimensionality of the physical reality get successful development (space-time physics, chronometrical theory and others), but nevertheless ideas connected with the World of events reality are not the riches of a wide circle of scientific interests. (A. D. Aleksandrov discussed some reasons for that in the paper [13].) It is a paradoxal fact for physics which must be founded on experimental data that the problem of direct experimental investigations of physical properties of the World of evens is not stated in general. For example, the problem of the corresponding *experimentum crucis* for the hypothesis on the reality of such physical object as the event, the study of possibilities of the natural interconnection of events through the temporal aspect, the analysis of proper temporal behaviour of the key characters of natural systems and processes and so on.

In this paper we present Kozyrev's point of view in the discussed question of instantaneous actions. In the next section Kozyrev's interpretation of instantaneous action-at-a-distance as the innate interconnections "by means of time" is given in the light of the World of events reality.

12.2 Innate interconnections of events with the null proper time interval

In order to see the physical meaning of Kozyrev's approach to the problem of instantaneous action-at-a-distance, it is of great importance to comprehend the reality of the World of events of the earthbound observer which manifests itself in some astrophysical observations of stars and stellar systems. Kozyrev's method of astronomical observations [14], [15] is distinguished by what light from celestial bodies is used mainly for orientation, see [16]. We can get the principal information when one of the two corresponding points of the celestial body's daily parallel is imaged onto a sensor of the measuring system by a mirror telescope: the one point is in the direction of the so-called true position of the celestial body and the other one is in the direction of the position which is symmetrical to the apparent position about the true one — this place is in fact the position of the body for the future, when a light signal sent now from the Earth would reach that body; we will call them the 'image-of-the-present' and the 'image-of-the-future', respectively.

Of course, it is an extraordinary occurrence for physics. That is why we must say that on the basis of this method N. A. Kozyrev suggested a new one of the determination of trigonometric stellar parallaxes and examined it for a lot of stars, trigonometric parallaxes of which were well known. The careful comparison of the experimental data Kozyrev obtained with those given in the General Catalogue [17] shows [14], [15] strong evidence for this new method, see [16]. Since such method was of considerable interest from a theoretical point of view, we repeated Kozyrev's observations for several stars. Our results [18], see [16], have confirmed those obtained by Kozyrev. (Later the corresponding results were also obtained in works of the Main Astronomical Observatory (Kiev) [19].) In rev. [16] Kozyrev's observations are considered in detail.

Just the coincidence of the trigonometric parallaxes of many stars (10UMa, αLeo, γBoo, εBoo, αLyr, ιPer, τPer, ζ^2Aqr, βPeg and others) obtained by this method of the direct measurement with their accepted values is the evidence that the corresponding prominent peak which can be found in the record of the sensor reaction to scanning of a star's daily parallel in the neighbourhood of the apparent position is the image-of-the-present peak.

Thus, there is the experimental fact of an instantaneous action-at-a-distance obtained with the aid of Kozyrev's method of astronomical observations by different authors, and the physical theory ought to comprehend it one way or another. Kozyrev's concept of the essence of time can be accepted or not, but, following physical conceptions of the space-time [7], physical ideas of the works [1], [20]–[22] and others, one cannot avoid the conclusion that astronomical observations by Kozyrev's method reveal in any case the possibility to study physical properties of the World of events reality. This will be evident at once if you take part, for example, in the observation of the three images for galaxy M31 (*Andromeda Nebula*): the image-of-the-past (with an accuracy to refraction coinciding with its apparent image), the image-of-the-present and the image-of-the-future [15]. (The author was lucky enough to look at these astronomical observations by N. A. Kozyrev in Crimean Astrophysical Observatory in 1980.) In the time of this observation, simultaneous visual control by sighting device (see Fig. 4 in [16]) does not single out any celestial object which would be projected on the sensor during the profile of the image-of-the-present and of the one of the image-of-the-future, too. The size of them along right ascension and along declination corresponds to the one of the galaxy M31. The angular distances both between the image-of-the-past and the image-of-the-present and between the image-of-the-present and the image-of-the-future were $(188 \pm 2)''$ along right ascension and $(34 \pm 2)''$ along declination that corresponds to the accepted data set on this galaxy. The angular distance between the image-of-the-past and the apparent image of *Andromeda Nebula* was of the order of $23''$ along declination that corresponds to the value of the refraction of the apparent position for that celestial object at the moment of observation: that value was equal to $23,6''$.

Another convincing instance: regular scanning of the celestial sphere by this method made it possible to reveal unknown large–scale objects on the celestial sphere, see the work [23].

Finally, our results of the solar experiment [24], see also [25], proposed by M. M. Lavrent'ev can be interpreted as the appearance of Sun's daily parallel as

Sun's World line if we consider the earthbound observer's celestial sphere as his World of events. Really, experimental research has revealed that a live system experiences an intensive superactivating influence not from the apparent Sun but from the place at $(8 \pm 1)^m$ ahead it along right ascension and the size of which along right ascension and along declination corresponds to the solar one — this "place" does be the event "true Sun" (if Sun's daily parallel is considered as Sun's World line), its temporal coordinate coincides with the instant of observation: the cells reaction to exposure at the focal plane of the telescope when the apparent Sun (note, the aperture of the telescope was closed by a plastic diaphragm in the course of the whole experiment) was projected onto it does not stand out against a "background" just as under the event true Sun the number of cells formed colonies increases strongly, see [25]. This effect of cells superactivation from the true Sun was observed also under the other tests: determination of spontaneous mutation background and determination of the cells ability for multiplication under unfavourable conditions — substrate starvation (organic compounds) without the temperature optimum for 10 hours. The other systems (some metal-film resistor and some geological specimen) experiences an intensive action (control over its resistance and weight, respectively) also from the event true Sun, see [25]. This interconnection of solar and terrain events looks like an instantaneous action-at-a-distance.

At the very beginning N. A. Kozyrev believed that there is the possibility of a connection of two space-time points not only through space (by means of corresponding quanta which are propagating with corresponding properties which are conditioned by characters of space), but through time as well. Later on, in the absence of refraction and other unusual properties of this kind of communication [10, pp. 363–383], [14], [15], see [16], he saw the fact of the absence of a material medium, as should be the case for this kind of interconnection — in time. His way of thinking is as follows. All the World is imaged onto the time axis of space-time as one point. It means that space does not have the size for time: time "transpierces" the space "all at once". Therefore, being based on his general views on nature of time, N. A. Kozyrev considers a distance-type action by means of time: a snap action of one object upon the other by any distance between them. In a mathematical sense, time is treated as a continuum of entities called instants or moments, which have a null length (duration) and are in the certain relations with themselves and with three-dimensional space, too, see Chap. 3, Sec. 7 in the monograph [26]. Accordingly, the snap action must be carried out through a null proper time interval $d\tau$. So that except for the instantaneous connection in a literal sense (when $dt = 0$, dt is the time interval in the frame of reference of an observer), the connection through one and the same moment of proper time may be accompanied by the motion over the surface of the light cone in the space-time in view of

$$d\tau = dt\sqrt{1 - v^2/c^2}, \qquad (1)$$

where v is a velocity at which the action is exerted, c is the speed of light.

Being based on his general views on nature of time, N. A. Kozyrev predicted the possibility to observe the true position of stars — the instantaneous action in a literal sense. When such possibility had been known experimentally also for the apparent position (with an accuracy to refraction: for this phenomenon refraction is absent) ($v = c$) and for the position which is symmetrical to the latter one about the true one ($v = -c$), N. A. Kozyrev turned his attention to Minkowskian space-time and began to explain the whole phenomenon from the view-point of the reality of the four-dimensional pseudo-Euclidean geometry [27].

It should be added that we have not to be surprised if instantaneous action-at-a-distance can look like some motion at the velocity of light in space: in this case, too, the deep interconnection of the temporal and spatial aspects of the World of events manifests itself.

12.3 External irreversible processes as an origin of instantaneous action-at-a-distance

In order to show what can be the physical origin of the instantaneous action in question, we use the results of Whitrow's approach [26] to the problem of the standard clock [7] in the light of Friedman's approach [5] to the problem of time. We consider this question in detail with the aim to demonstrate innate communication of the irreversible processes: this connection is not conditioned by phenomena in the space, for instance, by the propagation of definite quanta.

It is well-known that the physical concept of time is based on the definition, which supposes the use of some real process (motion) measured off equal durations. To mark out that process, we call it the 'fundamental' one [5]. Let us, as a fundamental process, some physical process λ is considered and, as a tentative duration measure, the change of its key property τ_λ is taken. (Such property has to change monotonously from the past to the future.)

We denote by $\tau_\lambda(i,j) = \tau_\lambda(i) - \tau_\lambda(j)$, where $\tau_\lambda(a)$ is the value of τ_λ at the moment of the event a, the measure of the duration between the event i and the subsequent event j. We shall suppose in what follows that all the events in question happen at the same point of space.

J. L. Synge [7] assumes that such duration measure makes no important physical sense and the concept of time will acquire greater concrete sense if we imply an existence of the *standard clock*, e.g., the atomic one. So that the concept of physical time is founded on the existence of special 'standard' processes. But how can base the choice of those standard processes? How can correctly, unambiguously introduce the definition of physical time (strictly speaking, duration)?

G. J. Whitrow [26] points out an important moment of the measurement of time: the measure of a duration which consists of two successive durations has to be equal to the arithmetical sum of their measures. That is because of the time additivity

$$t(i,j) + t(j,k) = t(i,k) \quad \forall i \prec j \prec k \tag{2}$$

where $t(a, b)$ is the actual measure of the duration between an event b and an event a preceded it, the sign '$\prec$' denotes 'is preceding'.

Generally speaking, if, as a duration measure, some measure τ_λ is taken

$$\tau_\lambda(i,k) \neq \tau_\lambda(i,j) + \tau_\lambda(j,k) \quad \forall i \prec j \prec k. \tag{3}$$

For example, the process λ is the radioactive decay and τ_λ is the part of decayed atoms.

That is why G. J. Whitrow introduces the concept of the temporal sum $\oplus$:

$$\tau_\lambda(i,k) = \tau_\lambda(i,j) \oplus \tau_\lambda(j,k) \quad \forall i \prec j \prec k,$$

that is determined with the aid of the following formula:

$$\tau_\lambda(i,j) \oplus \tau_\lambda(j,k) = \varphi_\lambda^{-1}\{\varphi_\lambda(\tau_\lambda(i,j)) + \varphi_\lambda(\tau_\lambda(j,k))\}, \tag{4}$$

where φ_λ is the definite monotonous function corresponding to the process λ, φ^{-1} is the inverse one. To obtain the (4), G. J. Whitrow used only the assumption of the general nature that the summation $\oplus$ of temporal quantities has to be associative and commutative.

Thus every process λ may be used in principle as the fundamental one with the aid of the measure

$$t(i,j) = \varphi_\lambda(\tau_\lambda(i,j)), \tag{5}$$

that satisfies the equation (2). It is the actual measure of the duration between two events i and j . As it is shown in ref. [26], the duration measure (5) is unique to within a scale factor. The standard clock is an instrument that can show the additive measure of durations (5).

Such an approach to the problem of the standard clock proves to be extremely helpful for mathematical interpretation of Kozyrev's conclusions about the external irreversible process: it can be a cause of some specific nonpower influence, which can change the matter state [28]. (Cf. this statement and I. Prigogine's position [29] on the irreversible processes.)

Really, because the physical time is universal, i.e. it is independent of 'clock mechanism', the (5) leads to the conclusion:

$$C_\lambda\varphi_\lambda(\tau_\lambda(i,j)) = C_\mu\varphi_\mu(\tau_\mu(i,j)) = C_\nu\varphi_\nu(\tau_\nu(i,j)) = ... \tag{6}$$

where C_λ, C_μ, C_ν,... are scale factors corresponding to the fundamental processes λ, μ, ν, ..., which are used as the 'clock mechanism'. Relation (6) means that these fundamental processes which look like independent ones proceed not independently: their "ways" intercommunicate in a special mode, that is conditioned by their *common existence in time*. It is not an action that "propagates in space", it is an *innate connection* "existent in time", in order words it is the instantaneous action, distance-type action. Moreover, it is clear, because of the deep interrelations of the spatial and temporal aspects of the World of events, this innate, temporal connection has to be accompanied by special motions in space, vide supra. This connection looks like the fact of change of the matter state of a system, including the rate of processes proceeding in it, because of the external irreversible process, i.e. like the fact of some influence of the external irreversible process, see [28].

The experimental researches of the above–mentioned influence, which was predicted and observed by Kozyrev, discovered a lot of its special properties, see [16]. The main ones are :

> — the presence of the two kinds of appearance of the influence in question, this kind is conditioned by the kind of the process himself (the latter is accompanied by growth or reduction of the entropy);
> — the possibility to determine the influence of stars and stellar systems, some geophysical processes upon the state of earthbound systems: physical, geological, biological.

In experiments of this type, we observed effects of the order of $10^{-5} \div 10^{-3}$, see [30], [31], [25] . We obtained several types of the reaction of the minerals weight upon some geophysical and stellar processes. Temporal behaviour of the systems key characters in connection with the influence of the external irreversible process is of the specific form. Note, the reaction of these systems to the change of the proper temperature conditions is quite different.

Our experimental experience allowed us to realize observations which used an extraordinary occurrence in the Solar system: the collision of the comet Shoemaker–Levy 9 with Jupiter in 1994. The observations were originally planned by way of the registration of a possible communication between some events on the Earth and the events of that catastrophe on Jupiter. Our observations [32] showed that the light signal was ≈ 43 minutes late than the reaction of some geological specimens and some metal-film resistor on the Earth to the corresponding events on Jupiter (impacts of fragments A, F, M, N, Q_2, Q_1 of the comet). The geocentric distance of Jupiter was 5.17804770 a.u. and light traveled from the position of that planet to the Earth in the course of 43.0 minutes. We suggested [32] to consider results of our measurements as the appearance of Jovian events being also based on many corresponding properties of this phenomenon.

12.4 Conclusions

Taking into account the aim of this collective work we sum up the main statements.

1. The temporal aspect of the World of events (space-time) is real as its spatial one. The spatial and temporal aspects are the two opposite, mutually complementing structures of existence of the real systems, phenomena and processes. That is why the interconnections of events which are realizing in the temporal aspect have not to be such as might be expected in the case of the spatial one.

The physical actions can be realized not only by means of the propagation of corresponding "quanta" in space at the corresponding velocity, but also by means of phenomena of innate connections in temporal aspect — in time. Their appearance looks like an instantaneous action-at-a-distance, there is not any propagation in space.

2. The innate interconnection of events has to realize for events with the null proper time interval. In accordance with formula (1) there are the three possibilities for the events connection of this kind. As a phenomenon in the temporal aspect, this interconnection has many corresponding properties. External irreversible processes are considered as the origin of the events interconnection in question.

3. The most interesting experimental results where this innate interconnection of events with the null proper time interval manifests itself were obtained in astrophysical observations of many stars (10UMa, αLeo, γBoo, εBoo, αLyr, ιPer, τPer, ζ^2Aqr, βPeg and others) and several stellar systems (galaxy M31 *Andromeda Nebula*, globular clusters M2 and M13) by N. A. Kozyrev's method. Three images of these celestial bodies were observed: the image-of-the-past, the image-of-the-present and the image-of-the-future. The image-of-the-past does not coincide with the apparent position of the celestial body if there is the visible refraction for the latter.

In connection with some papers authors of which ignore Kozyrev's approach to his observations, but believe that these observations support their own mathematical model well we have to say that interpretation of the three images of the celestial body is not the greatest proof of all. The main that is to be noted is Kozyrev's sensors of various nature — both their specific reaction and their corresponding physical properties which allow them to experience the influence of an external irreversible process (for instance, only certain type of resistors can be used as Kozyrev's sensor) — and also many physical circumstances of this phenomenon which were observed.

4. We consider the active influence observed from the region of Sun's daily parallel at 8^m ahead of the solar disk — it is the event "true Sun" if we consider the earthbound observer's celestial sphere as his World of events — as the appearance of the World of events reality. Together with the influence observed from the corresponding positions of stars and stellar systems for the future this phenomenon looks like the instantaneous action-at-a-distance.

In conclusion the author takes the opportunity to thank M. M. Lavrent'ev for his great interest and support of the work.

Bibliography

[1] Logunov A.A., *Lectures on the Theory of Relativity: Modern Analysis of the Problem* (Moscow University Press, Moscow, 1984) (in Russian).

[2] Poincaré H., "Sur la dynamique de l'électron" *Rendiconti del Circolo Matematico di Palermo*, **21**, 129 (1906).

[3] Logunov A.A., *To the Works by Henri Poincaré "On the Dynamics of the Electron"* (Moscow University Press, Moscow, 1988) (in Russian).

[4] Minkowski H., "Raum und Zeit" *Phys.Zs.*, **10**, 104 (1909).

[5] Friedman A.A., *The World as Space and Time*, 2nd Edition (Nauka, Moscow, 1965) (in Russian).

[6] Synge J.L., *Talking About Relativity* (North–Holland, Amsterdam, 1970).

[7] Synge J.L., *Relativity: The General Theory* (North–Holland, Amsterdam, 1960).

[8] Einstein A., *The Meaning of Relativity* (Princeton University Press, Princeton, N.Y., 1921).

[9] Synge J.L., "Introduction to General Relativity" *Relativity, Groups, and Topology* (Eds. C. DeWitt, B. DeWitt), 3–88 (Blackie and Son, London – Glasgow, 1963).

[10] Kozyrev N.A., *Selected Transactions* (Leningrad University Press, Leningrad, 1991) (in Russian).

[11] Newton I., *Philosophiae Naturalis Principia Mathematica* (London, 1871).

[12] Synge J.L., "A Plea for Chronometry" *The New Scientist*, **5**, 410 (1959).

[13] Aleksandrov A.D., "On the Philosophical Contents of Relativity" *Einstein and Philosophical Problems of Physics of the XX Century* (Ed. Eh.M. Chudinov), 117–137 (Nauka, Moscow, 1979) (in Russian).

[14] Kozyrev N.A. and Nasonov V.V., "A new method of determination of trigonometric parallaxes based on the measurement of the difference between the true and apparent star positions" *Astrometry and Celestial Mechanics* (Ed. V. K. Abalakin), 168–179 (Publication of the All-Union Astronomo-Geodesical Society of the USSR Acad. Sci., Moscow – Leningrad, 1978) (in Russian).

[15] Kozyrev N.A. and Nasonov V.V., "On some properties of time detected with the aid of astronomical observations" *Appearance of Cosmic Factors on the Earth and Stars* (Ed. V. S. Popov), 76–84 (Publication of the All-Union Astronomo-Geodesical Society of the USSR Acad. Sci., Moscow – Leningrad, 1980) (in Russian).

[16] Lavrent'ev M.M. and Eganova I.A., "Kozyrev's method of astronomical observations: information from true positions of stars, stellar systems, and planets" *Instantaneous Action at a Distance in Modern Physics: "Pro" and "Contra"* (Eds. A. E. Chubykalo, V. Pope, R. Smirnov-Rueda), 91–104 (Nova Science Publishers, Inc., Commack, New York, 1999).

[17] *General Catalogue of Trigonometric Stellar Parallaxes* (Yale University Press, New Haven, Conn., 1952).
Supplement to the General Catalogue of Trigonometric Stellar Parallaxes (Yale University Press, New Haven, Conn., 1963).

[18] Lavrent'ev M.M., Eganova I.A., Lutset M.K., and Fominykh S.F., "On the distance-type action of stars on a resistor" *Doklady AN SSSR*, **314**(2), 352–355 (1990) (in Russian).

[19] Akimov A.E., Koval'chuk G.U., Medvedev V.G., Olejnik V.K., Pugach A.F., "Preliminary Results of Astronomical Monitoring of Celestial Objects by N.A. Kozyrev's Method" *Preprint, Main Astron. Obs.* No GAO-92-5P (Kiev, 1992) (in Russian).

[20] Penrose R., *An Analysis of the Structure of Space-time* (Adam Prize Essay, Cambridge University, 1966).

[21] Segal I.,E., *Mathematical Cosmology and Extragalactic Astronomy* (Academic, N.Y., 1976).

[22] Burke W.L., *Spacetime, Geometry, Cosmology* (University Science Books, Mill Valley, California, 1980).

[23] Lavrent'ev M.M., Eganova I.A., Medvedev V.G., Olejnik V.K., and Fominykh S.F., "On scanning of the celestial sphere by Kozyrev's sensor" *Doklady AN SSSR*, **323**(4), 649–652 (1992) (in Russian).

[24] Lavrent'ev M.M., Gusev V.A., Eganova I.A., Lutset M.K., and Fominykh S.F., "On registration of the true Sun" *Doklady AN SSSR*, **315**(2), 368–370 (1990) (in Russian).

[25] Lavrent'ev M.M., "Interpretation of astronomical data connected with N. A. Kozyrev's discovery" *Memoirs on Differential Equations and Mathematical Physics*, **12**, 135–141 (1997).

[26] Whitrow G.J., *The Natural Philosophy of Time* (Nelson and Sons, London – Edinburgh, 1961).

[27] Kozyrev N.A., "Astronomical proof of the reality of the four-dimensional Minkowski geometry" *Appearance of Cosmic Factors on the Earth and Stars* (Ed. V. S. Popov), 85–93 (Publication of the All-Union Astronomo-Geodesical Society of the USSR Acad. Sci., Moscow – Leningrad, 1980) (in Russian).

[28] Kozyrev N.A.:

"Physical peculiarities of the components of double stars" *Colloque "On the Evolution of Double Stars", Uccle (Belgique), 29 Août – 2 Septembre 1966: Comptes Rendus/Union Astronomique Internationale (IAU)*, 197–202, 212, 252 (Communications/Observatoire Royal de Belgique; Sér. B, No. 17, 1967);

"On the possibility of experimental investigation of the properties of time" *Time in Science and Philosophy*, 111–132 (Prague, 1971);

"On the interaction between tectonic processes of the Earth and the Moon" *The Moon: [Proc. of] Symposium No. 47 of The International Astronomical Union Held at University of Newcastle-upon-Tyne, England, March 22–26, 1971*, 220–225 (Dordrecht-Holland, 1972).

[29] Prigogine I., *From Being to Becoming: Time and Complexity in the Physical Sciences* (W.H. Freeman and Company, San Francisco, 1980).

[30] Eganova I.A., *Analytical Survey of Ideas and Experiments in Modern Chronometry* (Dep. VINITI, No. 6423-84. Novosibirsk, 1984) (in Russian). Danchakov V.M. and Eganova I.A., *Microfield Experiments in Investigation of the Action of a Physical Irreversible Process* (Dep. VINITI, No. 8592-B87. Novosibirsk, 1987) (in Russian).

[31] Lavrent'ev M.M., Eganova I.A., Lutset M.K., Fominykh S.F., "On registration of the reaction of a substance to an external irreversible process" *Doklady AN SSSR*, **317**(3), 635-639 (1991) (in Russian).

[32] Lavrent'ev M.M., Eganova I.A., Gusev V.A., Borisov V.D., "To the Question about the Physical Reality of Space-time" *Proc. of the International Conference "Problem of Protecting the Earth from Collisions with Dangerous Cosmic Bodies"*, Part 2, 106–109 (Snezhinsk (Cheljabinsk-70), 1994).

In: Relativity, Gravitation, Cosmology
Editor: V. Dvoeglazov and A. Espinoza Garrido, pp. 163-192
ISBN 1-59033-981-9

Chapter 13

A Topological Perspective of Cosmology

R. M. Kiehn
Emeritus, Physics Dept., Univ. Houston
http://www.cartan.pair.com

Abstract

Methods of continuous topological evolution, expressed in terms of Cartan's theory of exterior differential forms, are used to construct a cosmological model of the present universe. The methods invoke topological, non-statistical, thermodynamic principles without geometric constraints of metric, connection or gauge. Stars and galaxies appear as self organizing topological defects, or condensates, of Pfaff topological dimension 3, embedded in a turbulent dissipative, but very dilute, non-equilibrium medium of Pfaff topological dimension 4. Such defects form long lived states of Pfaff dimension 3, and as such are states far from equilibrium which are of Pfaff topological dimension of 2 or less. The Jacobian matrix of the 1-form of Action used to describe the physical system leads to a universal phase function as a characteristic polynomial of fourth degree. The similarity invariants of the Jacobian matrix may be used to express the holomorphic polynomial in an intrinsic manner. The phase function can be interpreted as a family of implicit surfaces in an intrinsic 4D space, with a complex family, or order, parameter. The envelope of the family is homeomorphic to the Gibbs swallowtail surface of a van der Waals gas. The singular set of the family is homeomorphic to the equation of state of a van der Waals gas. The implication is that a domain of Pfaff topological dimension 4 can be deformed into a representation that mimics a van der Waals gas. If the gas is near its critical point, there are large fluctuations in density, which Landau has shown are correlated with a $1/r^2$ attractive force. Hence, a cosmological model based on the assumption that the universe is a dilute, non-equilibrium turbulent gas near its critical point explains the the granularity of the night sky, the inverse square law of gravitation, and the expansion of the universe.

13.1 Introduction

The objective of this article is to examine topological aspects and defects of thermodynamic physical systems and their possible continuous topological evolution, creation, and destruction on a cosmological scale. The creation and evolution of stars and galaxies will be interpreted herein in terms of the creation of topological defects and evolutionary phase changes in a very dilute turbulent, non-equilibrium, thermodynamic system of maximal Pfaff topological dimension 4. The cosmology so constructed is opposite in viewpoint to those efforts which hope to describe the universe in terms of properties inherent in the quantum world of Bose-Einstein condensates, super conductors, and superfluids [1]. Both approaches utilize the ideas of topological defects, but thermodynamically the approaches are opposite in the sense that the quantum method involves, essentially, equilibrium systems, while the approach presented herein is based upon non-equilibrium systems. Based upon the single assumption that the universe is a non-equilibrium thermodynamic system of Pfaff topological dimension 4 leads to a cosmology where the universe, at present, can be approximated in terms of the non-equilibrium states of a very dilute van der Waals gas near its critical point. The stars and the galaxies are the topological defects and coherent - but not equilibrium - structures of Pfaff topological dimension 3 in this non-equilibrium system of Pfaff topological dimension 4.

The motivation for this conjecture is based on the classical theory of correlations of fluctuations presented in the Landau-Lifshitz volume on statistical mechanics [3]. However, the methods used herein are not statistical, not quantum mechanical, and instead are based on Cartan's methods of exterior differential forms and their application to the topology of thermodynamic systems and their continuous topological evolution [4]. Landau and Lifshitz emphasize that real thermodynamic substances, near the thermodynamic critical point, exhibit extraordinary large fluctuations of density and entropy. In fact, these authors demonstrate that for an almost perfect gas near the critical point, the correlations of the fluctuations can be interpreted as a $1/\mathrm{r}$ potential giving a $1/\mathrm{r}^2$ force law of attraction. Hence, as a cosmological model, the almost perfect gas - such as a very dilute van der Waals gas - near the critical point yields a reason for both the granularity of the night sky and for the $1/\mathrm{r}^2$ force law ascribed to gravitational forces between for massive aggregates.

In this article, a topological (and non statistical) thermodynamic approach is used to demonstrate how a four dimensional variety can support a turbulent, non-equilibrium, physical system with universal properties that are homeomorphic (deformable) to a van der Waals gas. The method leads to the necessary conditions required for the existence, creation or destruction of topological defect structures in such a non-equilibrium system. For those physical systems that admit description in terms of an exterior differential 1-form of Action potentials of maximal rank, a Jacobian matrix can be generated in terms of the partial derivatives of the coefficient functions that define the 1-form of Action. When expressed in terms of intrinsic variables, known as the similarity invariants, the

Cayley-Hamilton 4 dimensional characteristic polynomial of the Jacobian matrix generates a universal phase function. Certain topological defect structures can be put into correspondence with constraints placed upon those (curvature) similarity invariants generated by the Cayley-Hamilton 4 dimensional characteristic polynomial. These constraints define equivalence classes of topological properties.

The characteristic polynomial, or Phase function, can be viewed as representing a family of implicit hypersurfaces. The hypersurface has an envelope which, when constrained to a minimal hypersurface, is related to a swallowtail bifurcation set. The swallowtail defect structure is homeomorphic to the Gibbs surface of a van der Waals gas. Another possible defect structure corresponds to the implicit hypersurface surface defined by a zero determinant condition imposed upon the Jacobian matrix. On 4 dimensional variety (space-time) , this non-degenerate hypersurface constraint leads to a cubic polynomial that always can be put into correspondence with a set of non-equilibrium thermodynamic states whose kernel is a van der Waals gas. Hence this universal topological method for creating a low density turbulent non-equilibrium media leads to the setting examined statistically by Landau and Lifschitz in terms of classical fluctuations about the critical point.

The conjecture presented herein is that non-equilibrium topological defects in a non-equilibrium 4 dimensional medium represent the stars and galaxies, which are gravitationally attracted singularities (correlations of fluctuations of density fluctuations) of a real gas near its critical point. Note that the Cartan methods do not impose (*a priori*) a constraint of a metric, connection, or gauge, but do utilize the topological properties associated with constraints placed on the similarity invariants of the universal phase function.

13.1.1 Topological Thermodynamics

The topological thermodynamic methods used herein are based upon Cartan's theory of exterior differential forms. The topological methods offer an understanding of the cosmos which is considerably different from the geometric approach assumed by the metrical theory of general relativity. The thermodynamic view assumes that the physical systems to be studied can be encoded in terms of a 1-form of Action Potentials, A, on a 4 dimensional variety of ordered independent variables, $\{\xi^1, \xi^2, \xi^3, \xi^4\}$. The variety supports a volume element $\Omega_4 = d\xi^1 \hat{}\, d\xi^2 \hat{}\, d\xi^3 \hat{}\, d\xi^4$. No metric, no connection, no constraint of gauge symmetry is imposed upon the 4 dimensional variety. Topological constraints will be imposed in terms of exterior differential systems [2]

In order to make the equations more suggestive to the reader, the symbolism for the variety of independent variables will be of the format $\{x, y, z, t\}$, but be aware that no constraints of metric or connection are imposed upon this variety. For instance, it is NOT assumed that the variety is euclidean. In that which follows another useful formalism of independent variables will be constructed in terms of the ordered set of similiarity invariant functions, which are given the symbols $\{X_M, Y_G, Z_A, T_K\}$. The similarity invariant functions are those deduced from the Jacobian matrix of the coefficients of that 1-form of Action, A, presumed to encode the properties of a physical system.

The 1-form of Action, A, will have components that form a covariant direction field, $A_k(x, y, z, t)$, to within a non-zero factor. Evolutionary processes will be determined in terms of 4 dimensional contravariant direction fields, $\mathbf{V}_4(x, y, z, t)$, to within a non-zero factor. Continuous topological evolution [4] will be defined in terms of Cartan's magic formula for the Lie differential, which, when acting on an exterior differential 1-form of Action, $A = A_k dx^k$, is equivalent *abstractly* to the first law of thermodynamics.

$$\begin{array}{lrcl} \text{Cartan's Magic Formula} & L_{(\mathbf{V}_4)}A & = & i(\mathbf{V}_4)dA + d(i(\mathbf{V}_4)A) \quad (13.1) \\ \text{First Law of Thermodynamics} & & : & W + dU = Q, \\ \text{Inexact 1-form of Heat} & L_{(\mathbf{V}_4)}A & = & Q \\ \text{Inexact 1-form of Work} & W & = & i(\mathbf{V}_4)dA, \\ \text{Internal Energy} & U & = & i(\mathbf{V}_4)A. \end{array}$$

In effect, Cartan's methods establish a topological basis of thermodynamics in terms of a theory of cohomology. The methods can be used to formulate precise mathematical definitions for many thermodynamic concepts in terms of topological properties -without the use of statistics or metric constraints. Moreover, the method applies to non-equilibrium thermodynamical systems and irreversible processes, again without the use of statistics or metric constraints.

The Pfaff Topological Dimension

One of the most useful topological tools is that defined as the Pfaff topological dimension. Recall that it is possible to define many topologies on the same set of elements. For any given exterior differential 1-form of functions, say $A = A_k(x, y, z, t)dx^k$, it is possible to construct the Pfaff sequence of terms, $\{A, dA, A\hat{}dA, dA\hat{}dA\}$. These elements may be used to construct a Cartan Topology (for any 1-form [5]). In the Cartan topology, the exterior derivative acts as limit point generator. Hence the union of a form and its exterior derivative create the topological closure of the form.

For any given 1-form, the Pfaff sequence will contain M successive non-zero terms equal to or less than N, the number of geometric dimensions of the base independent variables. The number M is defined as the "Pfaff topological dimension" or class of the given 1-form. The three important 1-forms of thermodynamics, A, W, and Q, can have different Pfaff dimensions. Suppose the 1-form of work is defined in terms of two functions as $W = PdV$. The Pfaff sequence consists of the terms $\{W, dW, 0, 0\}$; hence in this example, the Pfaff dimension of W is 2. From the first law, under the assumption that $W = PdV$,

$$\begin{aligned} Q &= W + dU = PdV + dU, \qquad (13.2) \\ dQ &= dW = dP\hat{}dV, \end{aligned}$$

$$Q\hat{}dQ = W\hat{}dW + dU\hat{}dW = 0 + dU\hat{}dP\hat{}dV$$
$$dQ\hat{}dQ = 0.$$

Hence, the Pfaff dimension of 2 for the work 1-form can be associated with a Pfaff dimension of 3 for the Heat 1-form, unless the Pressure is a function of the internal energy and the volume. In this latter case, the Pfaff dimension of Q and W are both 2.

In this article, attention will be focused on dissipative Turbulent systems with thermodynamic irreversible processes such that the Pfaff topological dimensions of A, W, and Q will be maximal and equal to 4. (The techniques can be extended to higher dimensional spaces.) These Turbulent systems of Pfaff dimension 4 are not topologically equivalent to Equilibrium systems (for which the topological dimension is 2, at most). Topological defects in the Turbulent state will be associated with sets of space time where the Pfaff topological dimensions are not maximal. It is remarkable that such topological defect sets can form attractors causing self organization and long lived states of Pfaff dimension 3, which are far from equilibrium.

Physical Systems

Isolated, Closed and Open Systems Physical systems and processes are elements of topological categories determined by the Pfaff topological dimension (or class) of the 1-forms of Action, A, Work, W, and Heat, Q. For example, the Pfaff topological dimension of the exterior differential 1-form of Action, A, determines the various species of thermodynamic systems in terms of distinct topological categories:

$$\begin{aligned} Systems &: \text{ defined by the Pfaff dimension of } A = \rho A^{(0)} \\ A\hat{}dA &= 0 \quad \text{Isolated - Pfaff dimension 2} \\ d(A\hat{}dA) &= 0 \quad \text{Closed - Pfaff dimension 3} \\ dA\hat{}dA &\neq 0. \quad \text{Open - Pfaff dimension 4.} \end{aligned} \tag{13.3}$$

In elementary thermodynamics it is often stated that isolated systems do not permit transport of energy or mass to the environment. Closed systems permit energy transport, but not mass transport to the environment. Open systems permit both energy and mass transport to the environment. Note that these topological specifications as given above are determined entirely from the functional properties of the physical system encoded as a 1-form of Action, A. The system topological categories do not involve a process, which is encoded by some vector direction field, $\mathbf{V}_4$. The cosmological model presented herein is based on an open, Pfaff dimension 4, non-equilibrium, Turbulent physical system, with internal defect structures of lesser Pfaff topological dimension acting as stars and galactic mass aggregates.

Equilibrium vs. Non-Equilibrium Systems The intuitive idea for an equilibrium system comes from the experimental recognition that the intensive variables of pressure and temperature become domain constants in an equilibrium state: $dP \Rightarrow 0$, $dT \Rightarrow 0$. A definition made herein is that the Pfaff dimension of a physical system in the equilibrium state is at most 2 [6]. The Cartan topology generated by the elements of the Pfaff sequence for A is then a connected topology of one component, $\{A \neq 0, dA \neq 0, A\hat{}dA = 0\}$. Although the Pfaff dimension of A is at most 2 in the equilibrium state, processes in the equilibrium state are such that the Work 1-form and the Heat 1-form must be of Pfaff dimension 1. For suppose $W = PdV$, then $dW = dP\hat{}dV \Rightarrow 0$ if the pressure is a domain constant. Similarly, suppose $Q = TdS$, the $dQ = dT\hat{}dS \Rightarrow 0$ if the temperature is a domain constant. Hence both W and Q are of Pfaff dimension 1 for this example.

A more stringent sufficient condition for equilibrium can be constructed in terms of the structure of the system, valid for any choice of process. For if the Pfaff dimension of the 1-form of Action is 1, then $dA \Rightarrow 0$. It follows that $W \Rightarrow 0$, hence the Pressure must vanish, and Heat 1-form is a perfect differential, $Q = d(U)$.

The cosmological model proposed herein is of Pfaff dimension 4, with defect structures of Pfaff dimension 3. Neither the dissipative system nor the defect structure is an "equilibrium" thermodynamic system, as the Pfaff dimension of such sets does not satisfy the criteria of equilibrium (where the Pfaff dimension is 2 or less). Although the defects in the Turbulent non-equilibrium regime are not necessarily equilibrium structures, once formed and self organized as coherent topological structures, they can evolve along extremal trajectories that are not dissipative, and may even have a Hamiltonian representation. These "stationary", if not long lived (excited) states, indeed are states far from equilibrium.

The descriptive words of self-organized states far from equilibrium are abstracted from the intuition and conjectures of I. Prigogine [7]. However, the topological theory presented herein presents for the first time a solid formal justification (with examples) for the Prigogine conjectures. Precise definitions of equilibrium and non-equilibrium systems, as well as reversible and irreversible processes can be made in terms of the topological features of Cartan's exterior calculus. Thermodynamic irreversibility and the arrow of time are well defined in a topological sense [8], a technique that goes beyond (and without) statistical analysis.

Multiple Components One of the most remarkable properties of the Cartan topology generated by a Pfaff sequence is due to the fact that when $A\hat{}dA = 0$, (Pfaff dimension 2 or less) the physical system is reducible to a single connected topological component. On the other hand when $A\hat{}dA \neq 0$, (Pfaff dimension 3 or more) the physical system admits more than one topological component. The bottom line is that when the Pfaff dimension is 3 or greater (such that conditions of the Frobenius unique integrability theorem are not satisfied), solution uniqueness to the Pfaffian differential equation, $A = 0$, is lost. If there exist solutions, there is more than one. Such concepts lead to propagating discontinuities (signals), envelope solutions (Huygen wavelets), an edge of regression and lack of time reversal invariance, and the existence of irreducible affine torsion in the theory of connections. It is the opinion of this author that a dogmatic insistence on uniqueness historically has hindered the understanding of irreversibility and non-equilibrium systems.

Processes

Reversible and Irreversible Processes. The Pfaff topological dimension of the exterior differential 1-form of Heat, Q, determines important topological categories of processes. From classical thermodynamics "The quantity of heat in a reversible process always has an integrating factor" [9] [10] . Hence, from the Frobenius unique integrability theorem, all reversible processes are such that the Pfaff dimension of Q is less than or equal to 2. Irreversible processes are such that the Pfaff dimension of Q is greater than 2. A dissipative irreversible topologically *turbulent* process is defined when the Pfaff dimension of Q is 4.

$$\begin{aligned} Processes &= \text{defined by the Pfaff dimension } Q \\ Q\hat{}dQ &= 0 \qquad \text{Reversible - Pfaff dimension 2} \\ d(Q\hat{}dQ) &\neq 0. \quad \text{Turbulent - Pfaff dimension 4.} \end{aligned} \tag{13.4}$$

Note that the Pfaff dimension of Q depends on both the choice of a process, $\mathbf{V}_4$, and the system, A, upon which it acts. As reversible thermodynamic processes are such that $Q\hat{}dQ = 0$, and irreversible thermodynamic processes are such that $Q\hat{}dQ \neq 0$, Cartan's formula of continuous topological evolution can be used to determine if a given process, $\mathbf{V}_4$, acting on a physical system, A, is thermodynamically reversible or not:

$$\begin{bmatrix} \text{Reversible Processes } \mathbf{V}_4 : \ L_{(\mathbf{V}_4)}A\hat{}L_{(\mathbf{V}_4)}dA = 0, \\ \text{Irreversible Processes } \mathbf{V}_4 : L_{(\mathbf{V}_4)}A\hat{}L_{(\mathbf{V}_4)}dA \neq 0. \end{bmatrix} \tag{13.5}$$

In this article it is assumed that the cosmological background for space-time belongs to the dissipative irreversible Turbulent non-equilibrium category, where the Pfaff topological dimension (or class) is maximal and equal to 4, almost everywhere, for each of the 1-forms of Action, A, Work, W, and Heat, Q. Of particular interest will be those subsets of space and time where the Turbulent non-equilibrium category admits, or evolves into, topological defects such that the Pfaff topological dimension for all three 1-forms is no longer maximal and equal to 4. Remarkably, Cartan's magic formula can be used to describe the continuous dynamic possibilities of both reversible and irreversible processes, in equilibrium or non-equilibrium systems, even when the evolution induces topological change, transitions between excited states, and changes of phase, such as condensations.

It is important to note that the velocity field need not be topologically constrained such that it is singularly parameterized. That is, the evolutionary processes described by Cartan's magic formula are not necessarily restricted to vector fields that satisfy the topological constraints of kinematic perfection, $dx^k - V^k dt = 0$. A discussion of topological fluctuations and an example fluctuation process is described in the last section.

Adiabatic Processes - Reversible and Irreversible The topological formulation permits a precise definition to be made for both reversible and an irreversible adiabatic processes in terms of the topological properties of Q. On a geometrical space of N dimensions, a 1-form will admit N-1 vector fields such that $i(V_A)Q = 0$. Such processes V_A are defined as adiabatic processes [6]. Note that adiabatic processes are defined by vector direction fields, to within an arbitrary factor, $\beta(x, y, z, t)$. That is, if $i(V_A)Q = 0$, then it is also true that $i(\beta V_A)Q = 0$. The differences between the inexact 1-forms of Work and Heat become obvious in terms of the topological format. Both 1-forms depend on the process and on the physical system. However, Work is always transversal to the process, as $i(\mathbf{V}_4)W = i(\mathbf{V}_4)i(\mathbf{V}_4)dA = 0$, but Heat is not, as $i(\mathbf{V}_4)Q = i(\mathbf{V}_4)dU \Rightarrow 0$, only for adiabatic processes.

It is not obvious that the adiabatic direction fields are such that the Pfaff dimension of Q is 2. That is, it is not obvious that Q can be written in the form, $Q = TdS$, as is possible on the manifold of equilibrium states. From the Cartan formulation it is apparent that if Q is not zero, then

$$\begin{aligned} i(\mathbf{V}_A)L_{(\mathbf{V}_A)}A &= i(\mathbf{V}_A)i(\mathbf{V}_A)dA + i(\mathbf{V}_A)d(i(\mathbf{V}_A)A) \\ &= 0 + i(\mathbf{V}_A)d(i(\mathbf{V}_A)A) = i(\mathbf{V}_A)Q \end{aligned} \quad (13.6)$$

Hence, for an Adiabatic process:

$$\text{Adiabatic process } 0 + i(\mathbf{V}_A)d(i(\mathbf{V}_A)A) = i(\mathbf{V}_A)Q \Rightarrow 0, \quad Q \neq 0. \quad (13.7)$$

A reversible process is defined such that Q is less than Pfaff dimension 3, or $Q\hat{}dQ = 0$ Hence $i(\mathbf{V}_A)(Q\hat{}dQ) = 0$. But

$$i(\mathbf{V}_A)(Q\hat{}dQ) = (i(\mathbf{V}_A)Q)\hat{}dQ - Q\hat{}i(\mathbf{V}_A)dQ \quad (13.8)$$

which permits reversible and irreversible adiabatic processes to be well defined[1] when $Q \neq 0$:

$$\begin{aligned} \text{Reversible Adiabatic Process} &= -Q\hat{}i(\mathbf{V}_A)dQ \Rightarrow 0, \; i(\mathbf{V}_A)Q \Rightarrow 0, \quad (13.9) \\ \text{Irreversible Adiabatic Process} &= -Q\hat{}i(\mathbf{V}_A)dQ \neq 0, \quad i(\mathbf{V}_A)Q \Rightarrow 0. \end{aligned}$$

[1]It is apparent that i(V)Q= 0 defines an adiabatic process, but not necessarily a reversible adiabatic process. This topological point clears up certain misconceptions that appear in the literature.

It is certainly true that if $L_{(\mathbf{V})}A = Q = 0$, *identically*, then all such processes are adiabatic, and reversible. In such cases, the Cartan formalism implies that $W + dU = 0$. Such systems are elements of the Hamiltonian class of processes, where $W = d\Theta$. Recall that *all* Hamiltonian processes are thermodynamically reversible. Hamiltonian processes are adiabatic when the internal energy $U = (i(V)A)$ is an evolutionary invariant.

$$\begin{aligned} \text{Hamiltonian Adiabatic Process} &= L_{(\mathbf{V})}\{i(V)A\} = i(V)Q = 0, \quad (13.10) \\ W &= i(V)dA = d\Theta, \\ i(V)W &= 0, \quad i(V)A = U. \end{aligned}$$

Note that for a given 1-form of heat, Q, it is possible to construct a matrix of N-1 null vectors, and then to compute the adjoint matrix of cofactors transposed to create the unique direction field (to within a factor), $\mathbf{V}_{NullAdjoint}$. Evolution in the direction of $\mathbf{V}_{NullAdjoint}$ does not represent an adiabatic process path, as $i(\mathbf{V}_{NullAdjoint})Q \neq 0$. For a given Q, the N-1 null vectors need not span a smooth hypersurface whose surface normal is proportional to a gradient field. The components of the 1-form may be viewed as the normal vector to an implicit hypersurface, but the implicit hypersurface is not necessarily defined as the zero set of some function.

13.1.2 Topological Torsion

For maximal, non-equilibrium, turbulent systems in space-time the maximal element in the Pfaff sequence generated by A, W, or Q, is a 4-form. On the geometric space of 4 independent variables, every 4-form is globally closed, in the sense that its exterior derivative vanishes everywhere. It follows that every 4-form is exact and can be generated by the exterior derivative of a 3-form. The exterior derivative of the 3-form is related to the concept of a divergence of a contravariant vector field. Most of the development in this article will be devoted to the study of such 3-forms, and their kernels. It is a remarkable fact that all 3-forms admit integrating denominators, such that their exterior derivative of a rescaled 3-form is zero almost everywhere. Space time points upon which the denominator has a zero value form defect topological structures.

When the Action for a physical system is of Pfaff dimension 4, there exists a unique direction field, $\mathbf{T}_4$, defined as the topological torsion 4-vector, that can be evaluated *entirely* in terms of those component functions of the 1-form of Action which define the physical system. To within a factor, this direction field[2] has the four components of the 3-form $A\hat{\ }dA$, with the properties such

[2] A direction field is defined by the components of a vector field which establish the "line of action" of the vector in a projective sense. An arbitrary factor times the direction field defines the same projective line of action, just reparameterized. In metric based situations, the arbitrary factor can be interpreted as a renormalization factor.

that

$$\begin{aligned} i(\mathbf{T}_4)\Omega_4 &= A\hat{}dA \\ W &= i(\mathbf{T}_4)dA = \sigma\ A, \\ U &= i(\mathbf{T}_4)A = 0, \\ Q\hat{}dQ &= L_{(\mathbf{T}_4)}A\hat{}L_{(\mathbf{T}_4)}dA = \sigma^2 A\hat{}dA \\ dA\hat{}dA &= 2\ \sigma\ \Omega_4. \end{aligned} \tag{13.11}$$

Hence, evolution in the direction of $\mathbf{T}_4$ is thermodynamically irreversible, when $\sigma \neq 0$ and A is of Pfaff dimension 4. The kernel of this vector field is defined as the zero set under the mapping induced by exterior differentiation. In engineering language, the kernel of this vector field are those point sets upon which the divergence of the vector field vanishes. The Pfaff dimension of the Action 1-form is 3 in the defect regions defined by the kernel of $\mathbf{T}_4$.

For purposes of more rapid comprehension, consider a 1-form of Action, A, with an exterior differential, dA, and a notation that admits an electromagnetic interpretation ($\mathbf{E} = -\partial\mathbf{A}/\partial t - \nabla\phi$, and $\mathbf{B} = \nabla \times \mathbf{A}$)[3]. The explicit format of $\mathbf{T}_4$ becomes:

$$\begin{aligned} \mathbf{T}_4 &= -[\mathbf{E}\times\mathbf{A}+\mathbf{B}\phi,\ \mathbf{A}\circ\mathbf{B}] \\ &\quad \text{Topological Torsion 4 vector,} \\ A\hat{}dA &= i(\mathbf{T}_4)\Omega_4 \\ &= T_4^x dy\hat{}dz\hat{}dt \\ &- T_4^y dx\hat{}dz\hat{}dt + T_4^z dx\hat{}dy\hat{}dt - T_4^t dx\hat{}dy\hat{}dz, \\ dA\hat{}dA &= 2(\mathbf{E}\circ\mathbf{B})\ \Omega_4 \\ &= \{\partial T_4^x/\partial x + \partial T_4^y/\partial y + \partial T_4^z/\partial z + \partial T_4^t/\partial t\}\ \Omega_4. \end{aligned} \tag{13.12}$$

When the divergence of the topological torsion vector is not zero, $\sigma = (\mathbf{E}\circ\mathbf{B}) \neq 0$, and A is of Pfaff dimension 4, W is of Pfaff dimension 4, and Q is of Pfaff dimension 4. The process generated by $\mathbf{T}_4$ is thermodynamically irreversible. The evolution of the volume element relative to the irreversible process $\mathbf{T}_4$ is given by the expression,

$$\begin{aligned} L(\mathbf{T}_4)\Omega_4 &= i(\mathbf{T}_4)d\Omega_4 + d(i(\mathbf{T}_4)\Omega_4) \\ &= 0 + d(A\hat{}dA) = 2(\mathbf{E}\circ\mathbf{B})\ \Omega_4. \end{aligned} \tag{13.13}$$

Hence, the differential volume element (and therefore the turbulent cosmological universe) is expanding or contracting depending on the sign and magnitude of $\mathbf{E}\circ\mathbf{B}$.

[3]The bold letter **A** represents the first 3 components of the 4 vector of potentials, with the order in agreement with the ordering of the independent variables. The letter A represents the 1-form of Action.

If A is (or becomes) of Pfaff dimension 3, then $dA\hat{}dA \Rightarrow 0$ which implies that $\sigma^2 \Rightarrow 0$, but $A\hat{}dA \neq 0$. The differential volume element Ω_4 is subsequently an evolutionary invariant, and evolution in the direction of the topological torsion vector is thermodynamically reversible. The physical system is not in equilibrium, but the divergence free $\mathbf{T}_4$ evolutionary process forces the Pfaff dimension of W to be zero, and the Pfaff dimension of Q to be at most 1. Indeed, a divergence free $\mathbf{T}_4$ evolutionary process has a Hamiltonian representation. In the domain of Pfaff dimension 3 for the Action, A, the subsequent continuous evolution of the system, A, relative to the process $\mathbf{T}_4$, proceeds in an energy conserving manner, representing a "stationary" or "excited" state far from equilibrium. These excited states can be interpreted as the evolutionary topological defects in the Turbulent dissipative system of Pfaff dimension 4.

On a geometric domain of 4 dimensions, assume that the evolutionary process generated by $\mathbf{T}_4$ starts from an initial condition (or state) where the Pfaff topological dimension of A is also 4. Depending on the sign of the divergence of $\mathbf{T}_4$, the process follows an irreversible path for which the divergence represents an expansion or a contraction. If the irreversible evolutionary path is attracted to a region (or state) where the Pfaff topological dimension of the 1-form of Action is 3, then $\mathbf{E} \circ \mathbf{B}$ becomes (or has decayed to) zero. The zero set of the function $\mathbf{E} \circ \mathbf{B}$ defines a hypersurface in the 4 dimensional space. If the process remains trapped on this hypersurface of Pfaff dimension 3, $\mathbf{E} \circ \mathbf{B}$ remains zero, and the $\mathbf{T}_4$ process becomes an extremal field. Such extremal fields are such that the virtual work 1-form vanishes, $W = i(\mathbf{T}_4)dA = 0$, and the now reversible $\mathbf{T}_4$ process has a Hamiltonian representation. The system is conservative in a Hamiltonian sense, but it is in a "excited" state on the hypersurface that is far from equilibrium, as the Pfaff dimension of the 1-form of Action is 3, and not 2. (If the path is attracted to a region where the function $\mathbf{E} \circ \mathbf{B}$ is oscillatory, the system evolutionary path defines a limit cycle.)

The fundamental claim made in this article is that it is these topological defects that self organize from the dissipative irreversible evolution of the Turbulent state into "stationary" states far from equilibrium that form the stars and the galaxies of the cosmos. They are the long lived remnants or wakes generated from irreversible processes in the dissipative non-equilibrium turbulent medium.

13.2 Thermodynamic Cosmology

13.2.1 The Jacobian Matrix of the Action 1-form.

The idea is to express the Jacobian matrix of the coefficient functions that define the 1-form if Action, A, in terms of "universal" coordinates. These universal coordinates will be the similarity invariants of the Jacobian matrix. For a 1-form of Action of Pfaff topological dimension 4, the Cayley-Hamilton theorem produces a Universal Phase function as a polynomial of 4th degree. What is remarkable about this Universal Phase function is that it has properties that are homeomorphically deformable into the format of a classic van der Waals gas. It is this universality that gives credence to the idea that the universe could be a non-equilibrium van der Waals gas near its critical point.

The Universal Characteristic Phase Function

The 1-form of Action, used to encode a physical system, contains other useful topological information, as well as geometric information. Consider the Turbulent thermodynamic state generated by a 1-form of Action, A, of Pfaff topological dimension 4. The component functions of the Action 1-form can be used to construct a 4x4 Jacobian matrix of partial derivatives, $[\mathbb{J}_{jk}] = [\partial(A)_j/\partial x^k]$. In general, this Jacobian matrix will be a 4 x 4 matrix that satisfies a 4th order Cayley-Hamilton characteristic polynomial $\Theta(x, y, z, t;\ \Psi)$ with 4 perhaps complex roots representing the perhaps complex eigenvalues, ρ_k, of the Jacobian matrix.

$$\Theta(x, y, z, t;\ \Psi) = \Psi^4 - X_M\Psi^3 + Y_G\Psi^2 - Z_A\Psi^1 + T_K \Rightarrow 0. \tag{13.14}$$

The functions $X_M(x, y, z, t), Y_G(x, y, z, t), Z_A(x, y, z, t), T_K(x, y, z, t)$ are the similarity invariants of the Jacobian matrix. If the eigenvalues are distinct, then the similarity invariants are given by the expressions:

$$\begin{aligned} X_M &= \rho_1 + \rho_2 + \rho_3 + \rho_4, \qquad (13.15)\\ Y_G &= \rho_1\rho_2 + \rho_2\rho_3 + \rho_3\rho_1 + \rho_4\rho_1 + \rho_4\rho_2 + \rho_4\rho_3,\\ Z_A &= \rho_1\rho_2\rho_3 + \rho_4\rho_1\rho_2 + \rho_4\rho_2\rho_3 + \rho_4\rho_3\rho_1,\\ T_K &= \rho_1\rho_2\rho_3\rho_4. \end{aligned}$$

The similarity invariants may be considered as a coordinate map from the original variety of independent variables, $\{x, y, z, t\} \Rightarrow \{X_M, Y_G, Z_A, T_K\}$. When the similarity invariants are treated as generalized coordinates, then the characteristic polynomial becomes a Universal Phase function, and will be used to encode universal thermodynamic properties.

Minimal surfaces

The Universal Phase function, Θ, may be considered as a family of hypersurfaces in the 4 dimensional space, $\{X_M, Y_G, Z_A, T_K\}$ with a complex family (order) parameter, Ψ. Moreover, it should be realized that the Universal Phase Function is a holomorphic function, $\Theta = \phi + i\chi$ in the complex variable $\Psi = u + iv$. That is

$$\Theta(X_M, Y_G, Z_A, T_K;\ \Psi) \Rightarrow \phi + i\chi, \tag{13.16}$$

where

$$\begin{aligned} \phi &= u^4 - 6u^2v^2 + v^4 - X_M(u^3 - 3uv^2) \qquad (13.17)\\ &+ Y_G(u^2 - v^2) - Z_A u + T_K\\ \chi &= 4u^3v - 4uv^3 - X_M(3u^2v - v^3) + 2Y_G uv - Z_A v. \end{aligned}$$

As such, in the 4D space of two complex variables, $\{\phi + i\chi, u + iv\}$, according to the theorem of Sophus Lie, any such holomorphic function produces a pair of conjugate *minimal* surfaces in the 4 dimensional space $\{\phi, \chi, u, v\}$. It follows that there exist a sequence of maps,

$$\{x, y, z, t\} \Rightarrow \{X_M, Y_G, Z_A, T_K\} \Rightarrow \{\phi, \chi, u, v\} \tag{13.18}$$

such that the family of hypersurfaces can be decomposed into a pair of conjugate minimal surface components. The criteria for a minimal surface is equivalent to the idea that $X_M = 0$. By suitable renormalization, the similarity invariant X_M is equivalent to the Mean Curvature of the hypersurface.

Envelopes

The theory of implicit hypersurfaces focuses attention upon the possibility that the Universal Phase function has an envelope. The existence of an envelope depends upon the possibility of finding a simultaneous solution to the two implicit surface equations of the family:

$$\Theta(x, y, z, t;\ \Psi) = \Psi^4 - X_M\Psi^3 + Y_G\Psi^2 - Z_A\Psi + T_K \Rightarrow 0. \tag{13.19}$$

$$\partial\Theta/\partial\Psi = \Theta_\Psi = 4\Psi^3 - 3X_M\Psi^2 + 2Y_G\Psi - Z_A \Rightarrow 0. \tag{13.20}$$

For the envelope to be smooth, it must be true that $\partial^2\Theta/\partial\Psi^2 = \Theta_{\Psi\Psi} \neq 0$, and that the exterior 2-form, $d\Theta\hat{}d\Theta_\Psi \neq 0$ subject to the constraint that the family parameter is a constant: $d\Psi = 0$. The envelope as a smooth hypersurface does not exist unless both conditions are satisfied. Recall that the envelope, if it exists, is a hypersurface in the space of similarity coordinates, $\{X_M, Y_G, Z_A, T_K\}$.

The envelope is determined by the discriminant of the Phase Function polynomial, which as a zero set is equal to a universal hypersurface in the 4 dimensional space of similarity variables $\{X_M, Y_G, Z_A, T_K\}$. This function can be written in terms of the similarity "coordinates" (suppressing the subscripts):

$$\begin{aligned}&\text{Discriminant of the Universal Phase Function} \\ =\ & 18X^3ZYT - 27Z^4 + Y^2X^2Z^2 - 4Y^3X^2T + 144YX^2T^2 \\ & +18XZ^3Y - 192XZT^2 - 6X^2Z^2T + 144TZ^2Y - 4X^3Z^3 \\ & -27X^4T^2 - 4Y^3Z^2 + 16Y^4T - 128Y^2T^2 + 256T^3 - 80XZY^2T.\end{aligned} \tag{13.21}$$

The discriminant has eliminated the family order parameter. Remarkably, choosing the constraint condition in terms of the dual condition that the Mean Curvature vanishes, $X_M \Rightarrow 0$, leads to a reduced discriminant, which defines a universal swallow tail surface homeomorphic (deformable) to the Gibbs surface of a van der Waals gas (subscripts suppressed):

$$\begin{aligned}&\text{Universal Gibbs Swallowtail Envelope } (X = 0, Y, Z, T) \\ =\ & -27Z^4 + 144TZ^2Y - 4Y^3Z^2 + 16Y^4T - 128Y^2T^2 + 256T^3 \Rightarrow 0.\end{aligned} \tag{13.22}$$

In other words, the Gibbs function for a van der Waals gas is a universal idea associated with minimal hypersurfaces, $X_K = 0$, of thermodynamic systems of Pfaff topological dimension 4. The similarity coordinate T_K plays the role of the Gibbs free energy, in terms of the Pressure $(\tilde{}Z_A)$ and the Temperature $(\tilde{}Y_G)$. The Spinodal line as a limit of phase stability, and the critical point are ideas that come from the study of a van der Waals gas, but herein it is apparent that these concepts are universal topological concepts that remain invariant with respect to deformations. The universal formulas for such constraints are presented in the next section. The result is that all thermodynamic systems of Pfaff topological dimension 4 are deformably equivalent to a van der Waals gas.

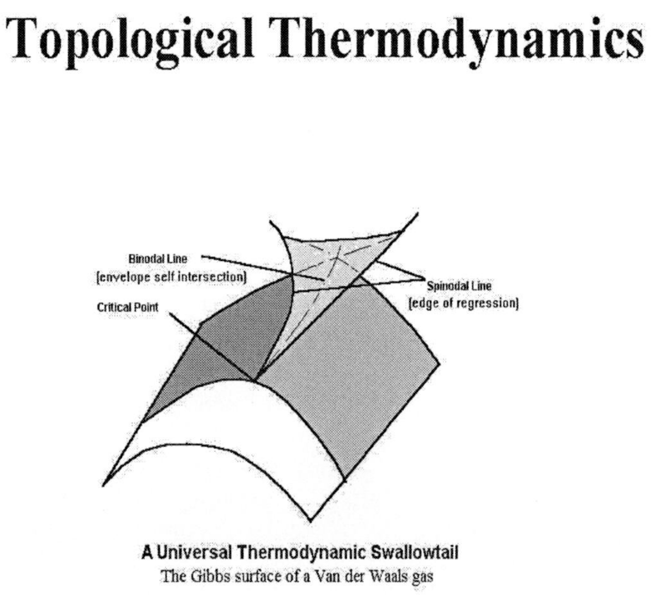

Figure 13.1:

It is important to recognize that the development of a universal non-equilibrium van der Waals gas has not utilized the concepts of metric, connection, statistics, relativity, gauge symmetries, or quantum mechanics.

The Edge of Regression and Self Intersections

The envelope is smooth as long as $\partial^2\Theta/\partial\Psi^2 = \Theta_{\Psi\Psi} \neq 0$, and that the exterior 2-form, $d\Theta\hat{}d\Theta_\Psi \neq 0$ subject to the constraint that the family parameter is a constant: $d\Psi = 0$. If $d\Theta\hat{}d\Theta_\Psi \neq 0$, but $\Theta_{\Psi\Psi} = 0$, then the envelope has a self intersection singularity. If $d\Theta\hat{}d\Theta_\Psi = 0$, but $\Theta_{\Psi\Psi} \neq 0$, there is no self intersection, and no envelope.

If the envelope exists, further singularities are determined by the higher order partial derivatives of the Universal Phase function with respect to Ψ.

$$\partial^2\Theta/\partial\Psi^2 = \Theta_{\Psi\Psi} = 12\Psi^2 - 6X_M\Psi + 2Y_G. \tag{13.23}$$

$$\partial^3\Theta/\partial\Psi^3 = \Theta_{\Psi\Psi\Psi} = 24\Psi - 6X_M \tag{13.24}$$

When $\partial^3\Theta/\partial\Psi^3 = \Theta_{\Psi\Psi\Psi} \neq 0$, and $d\Theta\hat{}d\Theta_\Psi\hat{}d\Theta_{\Psi\Psi} \neq 0$, the envelope terminates in a edge of regression. The edge of regression is determined by the simultaneous solution of $\Theta = 0, \Theta_\Psi = 0$ and $\Theta_{\Psi\Psi} = 0$. For the minimal surface representation of the Gibbs surface for a van der Waals gas, the edge of regression defines the Spinodal line of ultimate phase stability. The edge of regression is evident in the Swallowtail figure (Figure 2.1) describing the Gibbs function for a van der Waals gas.

If $\Theta_{\Psi\Psi\Psi} = 0$, then for $X_M = 0$, it follows that $Y_G = 0$, $Z_A = 0$, $T_K = 0$, which defines the critical point of the Gibbs function for the van der Waals gas. In other words, the critical point is the zero of the 4-dimensional space of similarity coordinates.

If $\Theta_{\Psi\Psi} = 0$, then for $X_M = 0$ the envelope has a self intersection. It follows from $\Theta_{\Psi\Psi} = 0$, that $\Psi^2 = -Y_G/6$, which when substituted into

$$\Theta_\Psi = 4\Psi^3 + 2Y_G\Psi - Z_A \Rightarrow 0, \tag{13.25}$$

yields the

$$\text{Universal Gibbs Edge of Regression: } Z_A^2 + Y_G^3(8/27) = 0, \tag{13.26}$$

which defines the Spinodal line, of the minimal surface representation for a universal non-equilibrium van der Waals gas, in terms of "similarity" coordinates.

Within the swallow tail region the "Gibbs" surface has 3 real roots and outside the swallow tail region there is a unique real root. The edge of regression furnished by the Cardano function defines the transition between real and imaginary root structures. The details of the universal non-equilibrium van der Waals gas in terms of envelopes and edges of regression with complex molal densities or order parameters will be presented elsewhere. These systems are not equilibrium systems for the Pfaff dimension is not 2. Of obvious importance is the idea that the a zero value for both Z_G and T_K are required to reduce the Pfaff dimension to 2, the necessary condition for an equilibrium system.

Ginsburg Landau Currents

The Universal Phase function can be solved for the determinant of the Jacobian matrix, which is equal to the similarity invariant T_K,

$$T_K = -\{\Psi^4 - X_M\Psi^3 + Y_G\Psi^2 - Z_A\Psi\}. \tag{13.27}$$

All determinants are in effect N - forms on the domain of independent variables. All N-forms can be related to the exterior derivative of some N-1 form or current, J. Hence

$$dJ = K\Omega_4 = div\mathbf{J} + \partial\rho/\partial t = -(\Psi^4 - X_M\Psi^3 + Y_G\Psi^2 - Z_A\Psi)\Omega_4. \qquad (13.28)$$

For currents of the form

$$\begin{aligned} \mathbf{J} &= grad\ \Psi, \\ \rho &= \Psi, \end{aligned} \qquad (13.29)$$

the Universal Phase function generates the universal Ginsburg Landau equations

$$\nabla^2\Psi + \partial\Psi/\partial t = -(\Psi^4 - X_M\Psi^3 + Y_G\Psi^2 - Z_A\Psi). \qquad (13.30)$$

13.2.2 Singularities as defects of Pfaff dimension 3

The family of hypersurfaces can be topologically constrained such that the topological dimension is reduced, and/or constraints can be imposed upon functions of the similarity variables forcing them to vanish. Such regions in the 4 dimensional topological domain indicate topological defects or thermodynamic changes of phase. It is remarkable that for a given 1-form of Action there are an infinite number rescaling functions, λ, such that the Jacobian matrix $\left[\mathbb{J}_{jk}^{scaled}\right] = [\partial(A/\lambda)_j/\partial x^k]$ is singular (has a zero determinant). For if the coefficients of any 1-form of Action are rescaled by a divisor generated by the Holder norm,

$$\text{Holder Norm:}\ \ \lambda = \{a(A_1)^p + b(A_2)^p + c(A_3)^p + e(A_4)^p\}^{m/p}, \qquad (13.31)$$

then the rescaled Jacobian matrix

$$\left[\mathbb{J}_{jk}^{scaled}\right] = [\partial(A/\lambda)_j/\partial x^k] \qquad (13.32)$$

will have a zero determinant, for any index p, any set of isotropy or signature constants, a, b, c, e, if the homogeneity index is equal to unity: $m = 1$. This homogeneous constraint implies that the similarity invariants become projective invariants, not just equi-affine invariants. Such species of topological defects can have the image of a 3-dimensional implicit characteristic hypersurface in space-time:

$$\text{Singular hypersurface in 4D:}\ \ \det[\partial(A/\lambda)_j/\partial x^k] \Rightarrow 0 \qquad (13.33)$$

The singular fourth order Cayley-Hamilton polynomial of $[\mathbb{J}_{jk}]$ then will have a cubic polynomial factor with one zero eigenvalue.

For example, consider the simple case where the determinant of the Jacobian vanishes: $T_K \Rightarrow 0$. Then the Phase function becomes

$$\begin{aligned}\text{Universal Equation of State} : \Theta(\{X_M, Y_G, Z_A, T_K = 0\};\ \Psi) \\ = \Psi(\Psi^3 - X_M\Psi^2 + Y_G\Psi - Z_A) \Rightarrow 0.\end{aligned} \qquad (13.34)$$

The space has been topologically reduced to 3 dimensions (one eigen value is zero), and the zero set of the resulting singular Universal Phase function becomes a universal cubic equation that is homeomorphic to the cubic equation of state for a van der Waals gas.

When the rescaling factor λ is chosen such that $p = 2, a = b = c = 1, m = 1$, then the Jacobian matrix, $[\mathbb{J}_{jk}]$, is equivalent to the "Shape" matrix for an implicit hypersurface in the theory of differential geometry. (See appendix 1.) Recall that the homogeneous similarity invariants can be put into correspondence with the linear Mean curvature, $X_M \Rightarrow C_M$, the quadratic Gauss curvature, $Y_G \Rightarrow C_G$, and the cubic Adjoint curvature, $Z_A \Rightarrow C_A$, of the hypersurface. The characteristic cubic polynomial can be put into correspondence with a nonlinear extension of an ideal gas *not necessarily* in an equilibrium state.

The Universal van der Waals gas

More that 100 years ago van der Waals introduced into the science of thermodynamics the equation of state now called the van der Waals gas:

$$P = \rho RT/(1 - b\rho) + a\rho^2 \qquad (13.35)$$

The van der Waals equation may be considered as a cubic constraint on the space of variables $\{n; P, V, T\}$ where $\rho = n/V$ is defined as the molar density.

$$\rho^3 - (1/b)\rho^2 + \{-(RT + bP)/ab\}\rho + P/ab = 0. \qquad (13.36)$$

This cubic equation is to be compared with the characteristic polynomial written in terms of the similarity invariants, M, G, and A. Note that the roots of the characteristic polynomial are not necessarily real. This observation leads to a well defined procedure for treating non-equilibrium thermodynamics systems as complex deviations from the real, or equilibrium, systems. The reality condition is determined by the Cardano function that describes an edge of regression discontinuity.

For a transformation such that

$$(8T + P)/3 = Y_G/(M/3)^2, \qquad (13.37)$$

$$\begin{aligned} P &= Z_K/(M/3)^3, \\ \lambda &= -\rho/(M/3), \end{aligned} \qquad (13.38)$$

the characteristic polynomial becomes an equation in terms of dimensionless parameters,

$$U(\lambda, T, P) = (\lambda)^3 - 3(\lambda)^2 + [(8T + P)/3](\lambda) - P = 0. \tag{13.39}$$

The last format given above is to be recognized as the Equation of State of a van der Waals Gas (compare to 13.34), in terms of dimensionless Pressure, Temperature relative to their values at the critical point.

13.2.3 The Falaco Cosmological Soliton

Although of importance to the cosmological concept of a universe expressible as a low density (non-equilibrium) van der Waals gas near its critical point, the factorization of the Jacobian characteristic polynomial into a cubic is not the only cosmological possibility. Of particular interest is the factorization that leads to a Hopf bifurcation. In this case the characteristic determinant vanishes, the Adjoint cubic curvature vanishes, the mean curvature vanishes (indicating a minimal surface), but the Gauss curvature is positive, and the two remaining eigenvalues of the characteristic polynomial are pure imaginary conjugates. Such results indicate rotations or oscillations (as in the chemical Brusselator reactions) and the possibility of spiral concentration or density waves on such minimal surfaces. Such structures at a cosmological level would appear to explain the origin of spiral arm galaxies. The Hopf type minimal surfaces of positive Gauss curvature do not represent thermodynamic equilibrium systems, for their curvatures, although two in number, are pure imaginary. The molal density distributions (or order parameters) are complex.

Evidence of such topological defects (at the macroscopic level) can be demonstrated by the creation of Falaco Solitons in a swimming pool [14]. These experiments demonstrate that such topological defects are available at all scales. The Falaco Solitons consist of spiral "vortex defect" structures (analogous to CGL theory) on a two dimensional minimal surface, one at each end of a 1-dimensional "vortex line" or thread (analogous to GPG theory). Remarkably the topological defect surface structure is locally unstable, as the surface is of negative Gauss curvature. Yet the pair of locally unstable 2-D surfaces is *globally* stabilized by the 1-D line defect attached to the "vertex" points of the minimal surfaces.

For some specific physical systems it can be demonstrated that period (circulation) integrals of the 1-form of Action potentials, A, lead to the concept of "vortex defect lines". The idea is extendable to "twisted vortex defect lines" in three dimensions. The "twisted vortex defects" become the spiral vortices of a Complex Ginsburg Landau (CGL) theory , while the "untwisted vortex lines" become the defects of Ginzburg-Pitaevskii-Gross (GPG) theory [13].

In the macroscopic domain, the experiments visually indicate "almost flat" spiral arm structures during the formative stages of the Falaco solitons. In the cosmological domain, it is suggested that these universal topological defects represent the ubiquitous "almost flat" spiral arm galaxies. Based on the experimental creation of Falaco Solitons in a swimming pool, it has been conjectured that M31 and the Milky Way galaxies could be connected by a topological defect thread [14]. Only recently has photographic evidence appeared suggesting that galaxies may be connected by strings.

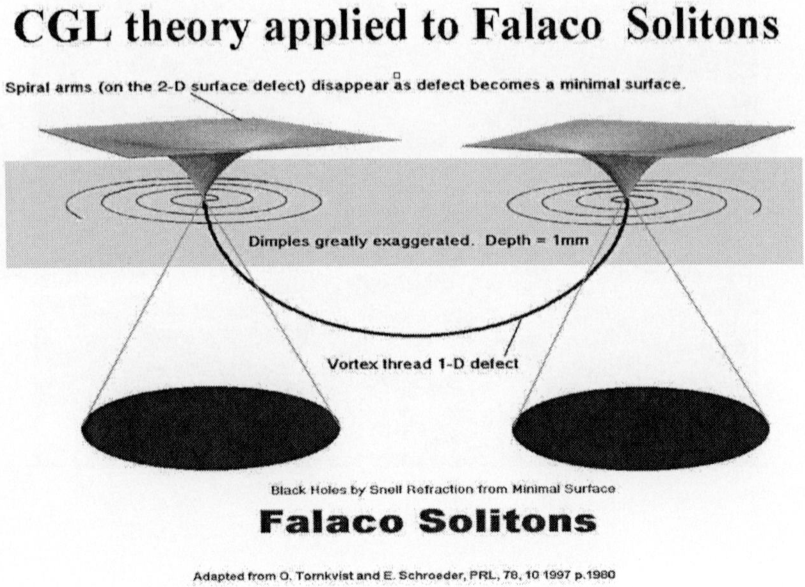

Figure 13.2:

At the other extreme, the rotational minimal surfaces of negative Gauss curvature which form the two endcaps of the Falaco soliton, like quarks, apparently are confined by the string. If the string (whose "tension" induces global stability of the unstable endcaps) is severed, the endcaps (like unconfined quarks in the elementary particle domain) disappear (in a non-diffusive manner). In the microscopic electromagnetic domain, the Falaco soliton structure offers an alternate, topological, pairing mechanism on a Fermi surface, that could serve as an alternate to the Cooper pairing in superconductors.

13.2.4 The Adjoint Current and Topological Spin

From the singular Jacobian matrix, $\left[\mathbb{J}_{jk}^{scaled}\right] = [\partial(A/\lambda)_j/\partial x^k]$, it is always possible to construct the Adjoint matrix as the matrix of cofactors transposed:

$$\text{Adjoint Matrix}: \left[\widehat{\mathbb{J}}^{kj}\right] = adjoint \left[\mathbb{J}_{jk}^{scaled}\right] \tag{13.40}$$

When this matrix is multiplied times the rescaled covector components, the result is the production of an adjoint current,

Cosmic Strings from Hubble ?

Figure 13.3:

$$\text{Adjoint current}: \left|\widehat{\mathbf{J}}^k\right\rangle = \left[\widehat{\mathbb{J}}^{kj}\right] \circ |\mathbf{A}_j/\lambda\rangle \tag{13.41}$$

It is remarkable that the construction is such that the Adjoint current 3-form, if not zero, has zero divergence globally:

$$\begin{aligned} \widehat{J} &= i(\widehat{\mathbf{J}}^k)\Omega_4 \\ d\widehat{J} &= 0. \end{aligned} \tag{13.42}$$

From the realization that the Adjoint matrix may admit a non-zero globally conserved 3-form density, or current, $\widehat{J}$, it follows abstractly that there exists a 2-form density of "excitations", $\widehat{G}$, such that

$$\text{Adjoint current}: \widehat{J} = d\widehat{G}. \tag{13.43}$$

$\widehat{G}$ is not uniquely defined in terms of the adjoint current, for $\widehat{G}$ could have closed components (gauge additions $\widehat{G}_c$, such that $d\widehat{G}_c = 0$), which do not contribute to the current, $\widehat{J}$.

From the topological theory of electromagnetism [11] [12] there exists a fundamental 3-form, $A\hat{}G$, defined as the "topological Spin" 3-form,

$$\text{Topological Spin 3-form} \ : \ A\hat{}G. \tag{13.44}$$

The exterior derivative of this 3-form produces a 4-form, with a coefficient energy density function that is composed of two parts:

$$d(A\hat{}G) = F\hat{}G - A\hat{}\widehat{J}. \tag{13.45}$$

The first term is twice the difference between the "magnetic" and the "electric" energy density, and is a factor of 2 times the Lagrangian usually chosen for the electromagnetic field in classic field theory:

$$\text{Lagrangian Field energy density} : F\hat{}G = \mathbf{B} \circ \mathbf{H} - \mathbf{D} \circ \mathbf{E} \tag{13.46}$$

The second term is defined as the "interaction energy density"

$$\text{Interaction energy density} : A\hat{}\widehat{J} = \mathbf{A} \circ \widehat{\mathbf{J}} - \rho\phi. \tag{13.47}$$

For the special (Gauss) choice of integrating denominator, λ with ($p = 2, a = b = c = 1, m = 1$) it can be demonstrated that the Jacobian similarity invariants are equal to the classic curvatures:

$$\{X_M, Y_G, Z_A, T_K\} \Rightarrow \{C_{M(mean\_linear)}, C_{G(gauss\_quadratic)}, C_{A(adjoint\_cubic)}, 0\}. \tag{13.48}$$

It can be demonstrated that the interaction density is exactly equal to the Adjoint curvature energy density:

$$\text{Interaction energy } A\hat{}\widehat{J} = C_A \; \Omega_4 \quad \text{(The Adjoint Cubic Curvature)}. \tag{13.49}$$

The conclusion reached is that a non-zero interaction energy density implies the thermodynamic system is not in an equilibrium state.

However, it is always possible to construct the 3-form, $\widehat{S}$:

$$\text{Topological Spin 3-form} : \widehat{S} = A\hat{}\widehat{G} \tag{13.50}$$

The exterior derivative of this 3-form leads to a co-homological structural equation similar the first law of thermodynamics, but useful for non-equilibrium systems. This result, now recognized as a statement applicable to non-equilibrium thermodynamic processes, was defined as the "Intrinsic Transport Theorem" in 1969 [15] :

$$\begin{aligned} \text{Intrinsic Transport Theorem (Spin)} \quad &: \quad d\widehat{S} = F\hat{}\widehat{G} - A\hat{}\widehat{J}, \qquad (13.51) \\ \text{First Law of Thermodynamics (Energy)} \quad &: \quad dU = Q - W \end{aligned}$$

If one considers a collapsing system, then the geometric curvatures increase with smaller scales. If Gauss quadratic curvature, C_G, is to be related to gravitational collapse of matter, then at some level of smaller scales a term cubic in curvatures, C_H, would dominate. It is conjectured that the cubic curvature produced by the interaction energy effect described above could prevent the collapse to a black hole. Cosmologists and relativists apparently have ignored such cubic curvature effects.

13.2.5 Topological Fluctuations

Topological fluctuations are admitted when the evolutionary vector direction fields are not singly parametrized:

$$\begin{aligned} \text{Fluctuations in position (pressure)} &: \quad d\mathbf{x} - \mathbf{v}dt = \Delta\mathbf{x} \neq 0 \qquad (13.52) \\ \text{Fluctuations in velocity (temperature)} &: \quad d\mathbf{v} - \mathbf{a}dt = \Delta\mathbf{v} \neq 0 \\ \text{Fluctuations in momenta (viscosity)} &: \quad d\mathbf{p} - \mathbf{f}dt = \Delta\mathbf{p} \neq 0. \end{aligned}$$

These failures of kinematic perfection undo the topological refinements imposed by a "kinematic particle" point of view, and place emphasis on the continuum methods inherent in fluids and plasmas. For example, consider the Cartan-Hilbert 1-form of Action on a space of 3n+1 independent variables (the p_μ are presumed to be independent Lagrange multipliers):

$$A = L(\mathbf{x}, \mathbf{v}, t)dt + p_\mu(dx^\mu - v^\mu dt) = L(x, v, t)dt + p_\mu \Delta x^\mu) \tag{13.53}$$

The Top Pfaffian in the Pfaff sequence is

$$(dA)^{n+1} = (n+1)!\{\Sigma_{\mu=1}^{n}(\partial L/\partial v^\mu - p_\mu)\bullet d\mathbf{v}^\mu\}\hat{}dp_1\hat{}...dp_n\hat{}dq^1\hat{}..dq^n\hat{}dt, \tag{13.54}$$

and yields a Pfaff dimension of 2n+2 for the 1-form of Action, defined on the geometric space of 3n+1 variables $\{x^\mu, p_\mu, v^\mu, t\}$. This even dimensional space defines a symplectic manifold.

For the maximal non-canonical symplectic physical system of Pfaff dimension 2n+2, consider evolutionary processes to be representable by vector fields of the form $\gamma V_{3n+1} = \gamma\{\mathbf{v}, \mathbf{a}, \mathbf{f}, 1\}$, relative to the independent variables $\{\mathbf{x}, \mathbf{v}, \mathbf{p}, t\}$. Define the "virtual work" 1-form, W, as $W = i(\mathbf{W})dA$, a 1-form which must vanish for the extremal case, and be non-zero, but closed, for the symplectic case. For any n, it may be shown by direct computation that the virtual work 1-form consists of two distinct terms, each involving a different fluctuation:

$$W = \{\mathbf{p} - \partial L/\partial\mathbf{v}\} \bullet \Delta\mathbf{v} + \{\mathbf{f} - \partial L/\partial\mathbf{x}\} \bullet \Delta\mathbf{x} \tag{13.55}$$

When the fluctuations in velocity are zero (temperature) and the fluctuations in position are zero (pressure), then the work 1-form will vanish, and the process and physical system admits a Hamiltonian representation. On the other hand if the fluctuations in velocity are not zero and the fluctuations in position are not zero, then the Work 1-form vanishes only if the momenta (the Lagrange multipliers, $\mathbf{p}$, are canonically defined ($\{\mathbf{p} - \partial L/\partial\mathbf{v}\} \Rightarrow 0$) and the Newtonian force is a gradient, $\{\mathbf{f} - \partial L/\partial\mathbf{x}\} \Rightarrow 0$. These topological constraints are ubiquitously assumed in classical mechanics.

When $\Delta x^k \Rightarrow 0$, such that all topological fluctuations vanish, then the Pfaff dimension of the physical system defined in terms of the Cartan-Hilbert 1-form of Action, A, is 2 (the equilibrium requirement).

13.3 Examples

In order to demonstrate content to the thermodynamic topological theory, two algebraically simple examples are presented below. The first corresponds to a Jacobian characteristic equation that has a cubic polynomial factor, and hence can be identified with a van der Waals gas. The second example exhibits the features associated with a Hopf bifurcation, where the characteristic equation has a quadratic factor with two pure imaginary roots, and two null roots. The third example demonstrates how a bowling ball, given initial angular momentum and energy, skids and/or slips changing its angular momentum and kinetic energy irreversibly via friction effects, until the dynamics is such that the ball rolls with out slipping. Once that "excited" state is reached, and topological fluctuations are ignored, the motion continues without dissipation. The system is in an excited state far from equilibrium.

Example 1: van der Waals properties from rotation and contraction

In this example,the Action 1-form is presumed to be of the form

$$A_0 = a(ydx - xdy) + b(tdz + zdt). \tag{13.56}$$

The 1-form of Potentials depends on the coefficients a and b. The results of the topological theory are (for $r^2 = x^2 + y^2 + z^2 + t^2$):

$$\begin{aligned}
\text{Mean curvature} &: \quad C_M = -2btz/(r^2)^{3/2} \qquad (13.57)\\
\text{Gauss curvature} &: \quad C_G = -\{b^2(x^2+y^2) - a^2(z^2+t^2)\}/(r^2)^2\\
\text{Adjoint curvature} &: \quad C_A = A\hat{}J_s = -2a^2btz/(r^2)^{5/2}\\
Top_Torsion &= \quad 2ab \cdot [0,0,z,-t]/(r^2)\\
\text{Adjoint Current} &: \quad J_s = (a^2b^2 \cdot [x,y,z,t])\,/(r^2)^2\\
\text{Pfaff Dimension 4} &: \quad dA\hat{}dA = 2ba(t^2 - z^2)/(r^2)^2\ \Omega_4
\end{aligned}$$

The Jacobian matrix has 1 zero eigen value and three non-zero eigenvalues. Hence, the cubic polynomial will yield an interpretation as a van der Waals gas. The Adjoint current represents a contraction in space-time, while the flow associated with the 1-form has a rotational component about the z axis.

Example 2: A Hopf 1-form

In this example,the Hopf 1-form is presumed to be of the form

$$A_0 = a(ydx - xdy) + b(tdz - zdt). \tag{13.58}$$

The 1-form of Potentials depends on the coefficients a and b. There are two cases corresponding to left and right handed "polarizations": $a = b$ or $a = -b$.

The results of the topological theory are (for $r^2 = x^2 + y^2 + z^2 + t^2$):

$$\begin{aligned} \text{Mean curvature} &: \quad C_M = 0, \\ \text{Gauss curvature} &: \quad C_G = \{b^2(x^2+y^2) \\ &+ \quad a^2(z^2+t^2)\}/(r^2)^2 \\ \text{Adjoint Cubic curvature} &: \quad C_A = A\hat{}J_s \; = 0 \\ Top\_Torsion &= \quad 2ab \; \cdot [x,y,z,t]/(r^2) \\ \text{Adjoint Current} &: \quad J_s = (ab/2) \; \cdot Top\_Torsion \\ \text{Pfaff Dimension 4} &: \quad dA\hat{}dA = 4ab/(r^2)\;\Omega_4 \end{aligned} \tag{13.59}$$

What is remarkable for this Action 1-form is that both the mean curvature and the Adjoint curvature of the implicit hypersurface in 4D vanish, for any choice of a or b. The Gauss curvature is non-zero, positive real and is equal to the inverse square of the radius of a 4D euclidean sphere, when $a^2 = b^2$. The Adjoint cubic interaction energy density is zero. The two non-zero curvatures are pure imaginary conjugates equal to

$$\rho = \pm\sqrt{-b^2(x^2+y^2) - a^2(z^2+t^2)}/(r^2). \tag{13.60}$$

The Hopf surface is a 2D imaginary *minimal* two dimensional hyper surface in 4D and has two non-zero imaginary curvatures! Strangely enough the charge-current density is not zero, but it is proportional to the Topological Torsion vector that generates the 3 form $A\hat{}F$. The topological Parity 4 form is not zero, and depends on the sign of the coefficients a and b. In other words the 'handedness' of the different 1-forms determines the orientation of the normal field with respect to the implicit surface. It is known that a process described by a vector proportional to the topological torsion vector in a domain where the topological parity is non-zero $4ba/(x^2 + y^2 + z^2 + t^2) \neq 0$ is thermodynamically irreversible.

The sliding bowling ball

Assume that the physical system of a bowling ball may be represented by a 1-form of Action constructed from a primitive Lagrange function with constraints. The Lagrange function is defined as

$$L = L(x, \theta, v, \omega, t) = \{\beta m(\lambda\omega)^2/2 - mv^2/2\} \tag{13.61}$$

Let the topological constraints be defined anholonomically by the Pfaffian system:

$$\{dx - vdt\} \Rightarrow 0, \qquad \{d\theta - \omega dt\} \Rightarrow 0, \;\; \{dx - \lambda d\theta\} \Rightarrow 0 \tag{13.62}$$

The constrained 1-form of Action becomes:

$$A = L(x, \theta, v, \omega, t)dt + p\{dx - vdt\} + l\{d\theta - \omega dt\} + s\{\lambda d\theta - dx\} \tag{13.63}$$

where $\{p, l, s\}$ are Lagrange multipliers. For simplicity, assume initially that two of the Lagrange multipliers (momenta) are defined canonically; e.g.,

$$p = \partial L/\partial v \Rightarrow -mv, \qquad l = \partial L/\partial \omega \Rightarrow \beta m\lambda^2\omega \tag{13.64}$$

which implies that

$$A = (-mv - s)dx + (\beta m\lambda^2\omega + \lambda s)d\theta - \{-mv^2/2 + \beta m(\lambda\omega)^2/2\}dt. \tag{13.65}$$

The Pfaff dimension of this action 1-form is 6. The volume element of the associated symplectic manifold is given by the expression

$$6Vol = 6m^2\beta\lambda^2\{v - \lambda\omega\}dx\hat{}d\theta\hat{}dv\hat{}d\omega\hat{}ds\hat{}dt = dA\hat{}dA\hat{}dA \tag{13.66}$$

The symplectic manifold has a singular subset upon which the Pfaff dimension of the Action 1-form is 2n+1 = 5. The constraint for such a contact manifold is precisely the no-slip condition:

$$\{v - \lambda\omega\} \Rightarrow 0. \tag{13.67}$$

On the 5 dimensional contact manifold there exists a unique extremal (Hamiltonian) field which (to within a projective factor) defines the conservative reversible part of the evolutionary process.

However, on the 6 dimensional symplectic manifold, there does not exist a unique extremal field, nor a unique stationary field, that can be used to define the dynamical system. There does exist a unique Topological Torsion direction field (or current) defined (to within a projective factor, σ) by the 6 components of the 5 form,

$$Torsion = A\hat{}dA\hat{}dA \tag{13.68}$$

This unique vector, T, has the properties that

$$L_{(\mathbf{T})}A = \Gamma \cdot A \quad and \quad i(\mathbf{T})A = 0. \tag{13.69}$$

This "Torsion" vector direction field satisfies the equation

$$L_{(\mathbf{T})}A\hat{}L_{(\mathbf{T})}dA = Q\hat{}dQ \neq 0. \tag{13.70}$$

Hence a process having a component constructed from this unique Torsion vector field becomes a candidate to describe the initial irreversible decay of angular momentum and kinetic energy.

Solving for the components of the Torsion vector for the bowling ball problem leads to the (unique) decaying dynamical system:

$$dv/dt = -\sigma/2\{\beta\lambda^2\omega^2 - 2\lambda v\omega + v^2\}/(v - \lambda\omega) \tag{13.71}$$

$$d\omega/dt = -\sigma/2\{-\beta\lambda^2\omega^2 + 2\beta\lambda v\omega - v^2\}/\lambda\beta(v - \lambda\omega) \quad (13.72)$$

$$ds/dt - \sigma s = \sigma/2\{-m\beta\lambda^2\omega^2 + mv^2\}/(v - \lambda\omega) \quad (13.73)$$

$$dx/dt = v \quad (13.74)$$

$$d\theta/dt = \omega \quad (13.75)$$

It is to be noted that the non-canonical "symplectic momentum" variables, defined by inspection from the constrained 1-form of Action as

$$P_x \doteq -(mv + s), \qquad P_\theta \doteq (m\beta\lambda^2\omega + s\lambda), \quad (13.76)$$

both decay irreversibly at the same (unit σ) rate on the manifold of Pfaff topological dimension 6, until the NoSlip condition is satisfied and the topology has become of Pfaff topological dimension 5.

Once the NoSlip condition is reached (irreversibly), the evolution proceeds without further topological change, in a conservative Hamiltonian manner.

13.4 Conclusions

Based upon the single assumption that the universe is a non-equilibrium thermodynamic system of Pfaff topological dimension 4 leads to a cosmology where the universe, at present, can be approximated in terms of the non-equilibrium states of a very dilute van der Waals gas near its critical point. The stars and the galaxies are the topological defects and coherent (but not equilibrium) self-organizing structures of Pfaff topological dimension 3 formed by irreversible topological evolution in this non-equilibrium system of Pfaff topological dimension 4.

The turbulent non-equilibrium thermodynamic cosmology of a real gas near its critical point yields an explanation for:

1. The granularity of the night sky as exhibited by stars and galaxies.
2. The Newtonian law of gravitational attraction proportional to $1/\mathrm{r}^2$.
3. The expansion of the universe.

13.5 Appendix 1. The shape matrix for implicit hypersurfaces.

Hypersurfaces can be defined in several ways. The most common is a differentiable parametric map from M dimensions into a variety of M+1 dimensions.

The surface is uniquely defined, with 1 component, and can be orientable or not orientable. A second way is to define the surface implicitly, in terms of the zero set of a function. Such surfaces are orientable, but can consist of multiple components. Some surfaces have both a parametric formulation and an implicit representation.

If in the parametric process the mapping is to a space constrained by a metric (not necessarily euclidean), then the M+1 x M+1 covariant metric on space M+1 has a well defined pullback, or preimage on the hypersurface, M. The induced M x M metric defines the first fundamental form of the surface. The more interesting properties are determined by the second fundamental form, or the shape matrix, which is related to properties of the normal field to the surface in M+1 space. Classical mboxs on differential geometry cover the parametric methods quite well.

However, the same cannot be said for implicit hypersurfaces, especially for the development of the equivalent of the shape matrix in terms of an implicit function representation. A somewhat obscure presentation is given by Harley Flanders. A more detailed presentation for implicit surface functions is given in O'Neill [16]. For surfaces that have both a parametric and an implicit representation, the present author discovered that there is an equivalence between the shape matrix, the Repere Mobile, and the Jacobian matrix of the surface normal field (the gradient of the implicit function) - if the normal field is scaled (divided) by the euclidean norm (the Gauss map). These results led to a more important formulation, where the previous results are extended to more general implicit surfaces, where the normal field is given, yet the normal field cannot be defined as a gradient of a single function. The normal field when expressed as a 1-form, need not be exact as it is in the classic theory of implicity surfaces. Indeed, the 1-form whose coefficients are form the normal field at a point can be of any Pfaff topological dimension in the general case, where it is always zero in the classic case.

The scaling of the normal field can also be done using the generalization of the euclidean distance given by the Holder norm 13.31. The key idea is that any form of the Holder norm which is homogeneous of degree 1 in the components of the normal field always leads to a Jacobian matrix with zero determinant. The constraint of a zero determinant forms a projective invariant system. Application to 4 dimensional spaces indicates a change of topological dimension as evolutionary properties evolve or are attracted to topological defect regions of zero determinant. The result is a domain of Pfaff dimension 3 occurring in regions of Pfaff topological dimension 4 [17]. For the Gauss map, the origin becomes a singular point, and is "removed" as a topological defect. For other versions of the Holder norm, the singular "points" have a variety of shapes. The theory of Homogeneous normal fields and the applications of the Holder norm will appear elsewhere.

Bibliography

[1] Volovik, G. E., *The Universe in a Helium droplet*, (Claredon Press, Oxford, 2003)

[2] Bryant, R.L.,Chern, S.S., Gardner, R.B.,Goldschmidt, H.L., and Griffiths, P. A. , *Exterior Differential Systems* (Springer, Verlag 1991)

[3] Landau, L. D. and Lifshitz, E. M., *Statistical Physics* (Pergamon Press, London 1958) p. 369.

[4] Kiehn, R.M. (1991) "Continuous Topological Evolution", *LANL arXiv/math-ph/0101032* (http://www22.pair.com/csdc/pdf/contevol3.pdf)

[5] Baldwin, P. and Kiehn, R. M. (1991) "Cartan's Topological Structure", Poster presented at the summer workshop, "Topology in Fluid Dynamics", *Inst. for Th. Phy.* UCSB, August 1991. arXiv math-ph/0101033, http://www22.pair.com/csdc/pdf/topstru4.pdf

[6] Bamberg, P. and Sternberg, S., *A course in mathematics for students of physics, 2*, (Cambridge University press, Cambridge, 1992) p. 775- 780.

[7] Kondepudi, D. and Prigogine, I., *Modern Thermodynamics* (Wiley, New York, 1998).

[8] Kiehn, R. M. (2003), "Thermodynamic Irreversibility and the Arrow of Time" in R. Bucher et al. (eds.), *The Nature of Time: Geometry, Physics and Perception*, 243-250. (http://www22.pair.com/csdc/pdf/arwfinal.pdf).

[9] Goldenblat, I.I., "Some problems of the mechanics of continuous media", (Noordhoff, Holland 1962) p.193 *The quantity of heat in a reversible process always has an integrating factor.*

[10] Morse, P.M., *Thermal Physics*, (Benjamin, NY, 1964) p. 60.

[11] Kiehn, R. M., "A topological perspective of Electromagnetism", to be published in *Classical electrodynamics: new horizons*, Editors Andrew E. Chubykalo and Roman Smirnov-Rueda, Rinton Press, Inc. (USA). Also see http://www22.pair.com/csdc/pdf/rmktop.pdf.

[12] Kiehn, R. M. (1999), "Topological evolution of classical electromagnetic fields and the photon" in Dvoeglazov, V., ed, *Photon and Poincaré Group,* Nova Science Publishers, Inc., Commack, New York, 1999, p.246-262. ISBN 1-56072-718-7 (http://www22.pair.com/csdc/pdf/photon.pdf).

[13] Tornkvist, O. and Schroeder, E., (1997) "Vortex dynamics in Dissipative Systems", *Phys Rev. Lett.* 78, 10 , p. 1908-1911.

[14] Kiehn, R. M., (1987) The Falaco Effect as a topological defect was first noticed by the present author in the swimming pool of an old MIT friend, during a visit in Rio de Janeiro, at the time of Halley's comet, March 1986. The concept was presented at the Austin Meeting of Dynamic Days in Austin, January 1987, and caused some interest among the resident topologists. The easily reproduced experiment added to the credence of topological defects in fluids. It is now perceived that this topological phenomena is universal, and will appear at all levels from the microscopic to the galactic. *LANL arXiv/gr-qc/0101098* (http://www22.pair.com/csdc/pdf/falaco97.pdf).

[15] Kiehn, R.M. and Pierce, J. F., (1969) "An Intrinsic Transport Theorem." *Phys. Fluids*, 12, p.1971. (http://www22.pair.com/csdc/pdf/inttrans.pdf)

[16] Flanders, H. (1963) *Differential Forms* (Academic Press, New York) p.126. O'Neill, B. *Elementary Differential Geometry* (Academic Press, New York) p.189...

[17] *http://www22.pair.com/csdc/pdf/implinor.pdf* and *http://www22.pair.com/csdc/pdf/parametric.pdf*

Index